ILLUSTRATED INVERTEBRATE ANATOMY

ILLUSTRATED INVERTEBRATE ANATOMY

A Laboratory Guide

Sidney K. Pierce and Timothy K. Maugel
with original illustrations by Lois Reid

New York   Oxford

OXFORD UNIVERSITY PRESS

1987

Oxford University Press

Oxford   New York   Toronto
Delhi   Bombay   Calcutta   Madras   Karachi
Petaling Jaya   Singapore   Hong Kong   Tokyo
Nairobi   Dar es Salaam   Cape Town
Melbourne   Auckland

and associated companies in
Beirut   Berlin   Ibadan   Nicosia

Published by Oxford University Press, Inc.,
200 Madison Avenue, New York, New York 10016

Oxford is a registered trademark of Oxford University Press

Library of Congress Cataloging-in-Publication Data
Pierce, Sidney K.
Illustrated invertebrate anatomy.
Includes index.
1. Invertebrates—Anatomy—Laboratory manuals. I. Maugel, Timothy K.
II. Reid, Lois.   III. Title. QL363.P54   1987   592′.04   86-5408
ISBN 0-19-504071-6
ISBN 0-19-504077-5 (pbk.)

Printing (Last digit): 9 8 7 6 5 4 3 2 1

Printed in the United States of America
on acid-free paper

To Chris, Alisa, Mike, Connie, and Nicole,
who put up with us as we proceeded.

PREFACE

A glance through the biological literature indicates that a wide array of invertebrate species is being investigated by people with interests ranging from field ecology to subcellular biology. Similarly, academic courses in invertebrate zoology, once taught by invertebrate zoologists *in sensu stricto,* are now offered by scientists who are often experts in areas quite far afield from classical invertebrate zoology. As a result, the approaches to and material covered in invertebrate zoology courses are diverse. This diversity is not a problem for the lecture portion of these courses, for there are texts available that easily supplement the lecture material.

The laboratory portion of an invertebrate zoology course is another matter altogether. Several laboratory manuals are available that offer a variety of types of laboratory exercises, but in our experience many of the exercises are not appropriate for every laboratory setting. In fact, most laboratory courses consist of a collection of the instructor's favorite experiments, designed with the constraints of a particular class in mind: available species, animal-holding facilities, and instrumentation. Thus, many instructors come to rely on their own sets of exercises. What is needed to make the teaching laboratory a success is a good manual of invertebrate anatomy to supplement the exercises.

There are several sources of anatomical information. Traditional laboratory manuals provide at least some anatomy, often in idealized drawings. These illustrations are useful, but they do not always accurately reflect what is available in the laboratory. There are also a few dissection manuals, some excellent but limited in scope (for example, F. A. Brown's classical work *Selected Invertebrate Types,* now out of print), others too massive and all-encompassing for routine student use (for example, Libbie Hyman's monumental effort, *The Invertebrates,* in six volumes). There seems to be a place for a laboratory book that presents both photographs and accurate anatomical drawings of the species that students are likely to encounter in a teaching laboratory in invertebrate zoology—whatever the instructor's approach.

We set out to assemble such a pictorial collection several years ago and it is presented on the following pages. We have made an effort to include only those species that are readily available, either alive or preserved, from scientific supply houses in North America. We have not included insects or, in general, parasites, as these groups are often treated in separate courses. The photographs are not intended to cover every anatomical detail but instead represent what the student is likely to encounter in the teaching laboratory. We have supplemented the dissections and light micrographs with substantial numbers of scanning electron micrographs, which should help the students visualize and better understand the anatomical features. Finally, we have kept the text to a minimum. The figure captions do not exhaustively cover the details of each figure. Instead, they supplement the anatomy, briefly describing the areas of special importance or interest for each specimen.

We are grateful to the many people who contributed to this book. Without them, it would have taken us much longer to complete our task and the book would not be what it is. Most of the drawings were done by Lois Reid. Along with contributing her obvious artistic talents, Ms. Reid spent long hours learning anatomy to ensure the accuracy of her drawings. We are especially grateful for her meticulous work and her dedication to the project. Another artist, S. Brust, contributed several drawings during the early stages of the project. We are also grateful to everyone who permitted us to use original photographs or who supplied specimens. Many of these contributors are cited in the figure captions. In addition, G. Schumacher, K. Bart, D. W. Coats, T. Sawyer, A. Barnett, and S. Suchard contributed to the Protista section; R. Blanquet contributed to the Cnidaria section; S. Coon, C. Sullivan, and Rita contributed to the Mollusca section; and A. Pierce contributed to the Arthropoda section. Eileen Barnett spent long hours typing the text. Dave Hinkle photographed most of the artwork. Finally, we are grateful to the animal collectors of the Marine Resources Department of the Marine Biological Laboratory at Woods Hole who provided us with excellent specimens, often on short notice.

This book is contribution no. 240 from the Tallahassee, Sopchoppy and Gulf Coast Marine Biological Association, Inc. It is also contribution no. 45 from the Laboratory for Ultrastructural Research of the University of Maryland.

Woods Hole   S.K.P.
1985   T.K.M.

CONTENTS

ILLUSTRATED INVERTEBRATE ANATOMY

SECTION I
Taxonomic Summary

KINGDOM PROTISTA
Phylum Mastigophora
  CLASS PHYTOMASTIGOPHOREA
    Order Euglenida
      Euglena
      Peranema
    Order Volvocida
      Chlamydomonas
      Gonium
      Pandorina
      Eudorina
      Volvox
    Order Dinoflagellida
      Ceratium
      Peridinium
      Gyrodinium
      Ornithocercus
Phylum Sarcodina
  CLASS RHIZOPODEA
    Order Amoebida
      Amoeba
    Order Testacisa (Arcellinida)
      Arcella
      Difflugia
    Order Foraminiferida
  CLASS ACTINOPODEA
  SUBCLASS RADIOLARIA
  SUBCLASS HELIOZOA
      Actinosphaerium
Phylum Ciliophora
  CLASS KINETOFRAGMINOPHORA
  SUBCLASS GYMNOSTOMATA
      Didinium
  CLASS OLIGOHYMENOPHORA
  SUBCLASS HYMENOSTOMATA
      Paramecium
      Tetrahymena
  SUBCLASS PERITRICHA
      Carchesium
      Epistylis
      Vorticella
      Zoothamnium
  CLASS POLYHYMENOPHORA
  SUBCLASS SPIROTRICHA
    Order Heterotrichida
      Stentor
    Order Hypotrichida
      Euplotes

FIG. 1. Light micrograph (LM, $\times 2200$) of the green flagellate *Euglena gracilis.* The green color in living *Euglena* is due to the presence of chlorophyll in chloroplasts, where photosynthesis takes place. The carbohydrate produced is stored as the paramylon bodies. Several euglenoid species have this plantlike characteristic, but all require a carbon source more complex than CO_2. Some *Euglena* species lose their chloroplasts if they are cultured in the dark, thus becoming colorless and requiring more complex carbon and nitrogen sources for metabolism.

FIG. 2. Scanning electron micrographs (SEMs) of *Euglena* showing external morphology. The outer surface, the pellicle, is sculptured into grooves that run helically around the longitudinal axis of the cell (Fig. 2a, $\times 1700$). The grooves are produced by pellicular strips, which are structures composed of fibrous protein and microtubules, underlying the cell membrane. The pellicular grooves are shown in close-up in SEM (Fig. 2b, $\times 12,000$). *(Photograph by E. B. Small, University of Maryland.)* The long flagellum, used for locomotion, is not smooth; its surface is a mesh of fine fibrils called mastigoneme processes (Fig. 2c, $\times 14,000$).

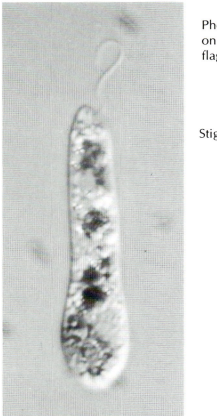
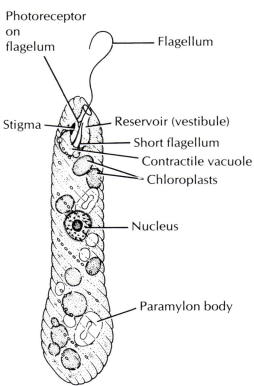

Photoreceptor
on
flagelum

Flagellum

Stigma

Reservoir (vestibule)

Short flagellum

Contractile vacuole

Chloroplasts

Nucleus

Paramylon body

FIG. 1

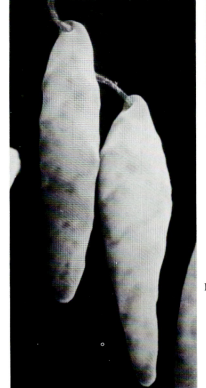

FIG. 2a

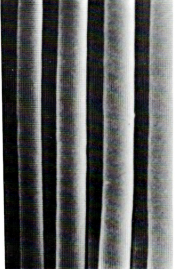

FIG. 2b

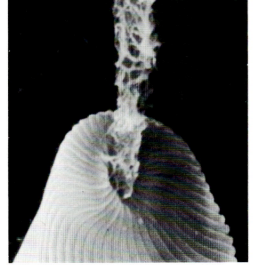

FIG. 2c

FIGS. 3 and 4. Although many species of *Euglena* have plantlike nutritive characteristics, other euglenoids are animal-like in their feeding mode. *Peranema,* the colorless euglenoid flagellate shown here in LM (Fig. 3, ×1560) and SEM (Fig. 4, ×2400), is an active carnivore. It feeds on a variety of bacterial and protozoan species that it ingests using two parallel structures, collectively called the rod organ, that lie in the cytoplasm next to the reservoir. Although many accounts indicate that the rod organ is protruded through the permanent cytostome and either attaches to the prey or is used to puncture the prey cell, recent studies indicate otherwise. The role of the rod organ in feeding seems to be only mechanical, supporting and helping expand the cytostomal opening. The cytostome, which in *Peranema* opens next to the entrance of the reservoir, becomes distended and the prey is engulfed into a food vacuole. Paramylon bodies are present in *Peranema,* but their source may be exogenous since chloroplasts are not present.

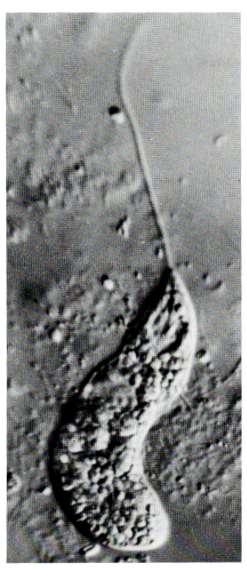

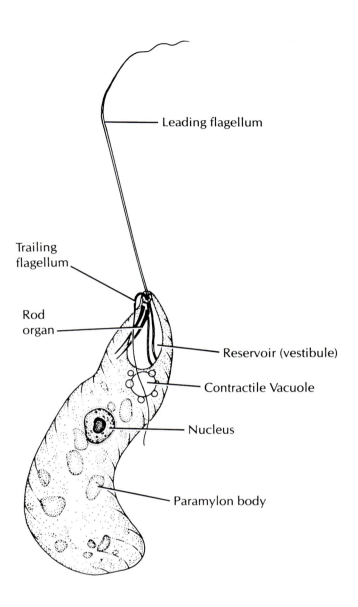

Leading flagellum

Trailing flagellum

Rod organ

Reservoir (vestibule)

Contractile Vacuole

Nucleus

Paramylon body

FIG. 3

FIG. 4

FIGS. 5–9. Over the years, several theories have been put forward to explain the process by which multicellular organisms have evolved from unicellular ones. One of the more persistent of these theories, proposed more than a century ago by Haeckel and later given legitimacy by both Metschnikoff and Hyman, is called the colonial theory. It proposes that the metazoa arose from colonial flagellates, using both the occurrences of flagellated sperm cells in the lower metazoa and true egg and sperm cells in the phytoflagellates as evidence. In addition, there is a tendency toward cellular specialization into somatic and reproductive cells in some of the colonial flagellates, which might be construed as a harbinger of multicellularity.

The colonial theory has had many adherents over the years, as well as many critics. The colonial flagellates have characteristics not represented in the metazoa and are therefore bothersome to the theory's detractors. For example, the plantlike traits in the Volvocids—cell walls and chloroplasts—the occurrence of meiotic reduction division after fertilization, their freshwater habitat, and even the evolutionary period required to produce an organism as highly specialized as *Volvox* all argue against the likelihood that the origins of the metazoa reside within the Volvocids.

The species shown in Figs. 5–9, all from the order Volvocida, represent a model progression from single-cell to multicellular organism. *Chlamydomonas* (Fig. 5, ×1880) carries out all cellular functions, including reproduction, as a single cell. *Gonium* (Fig. 6, ×400) consists of four cells held together by a gelatinous matrix. Each cell is flagellated and performs all somatic and reproductive functions. *Pandorina* (Fig. 7, ×450) is a sixteen-cell sphere, and *Eudorina* (Fig. 8, ×600) is a thirty-two-cell sphere. All the cells of these larger colonies are still capable of both asexual and sexual reproduction. The colonies are polarized, always swimming in one direction. The zenith of the colonial form is represented by *Volvox* (Fig. 9, ×130), which can consist of hundreds, sometimes thousands, of cells. The majority of cells in the colony cannot reproduce. The reproductive cells are larger than the somatic cells and are located toward the posterior pole of the colony (the region which is always aft when the colony swims). The *Volvox* colony shown in Fig. 9 is full of daughter colonies that will be released when the parent colony ruptures. Figs. 10–13 show *Volvox* in more detail.

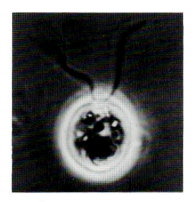

FIG. 5

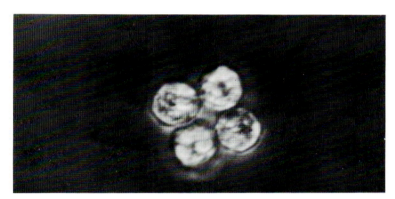

FIG. 6

FIG. 7

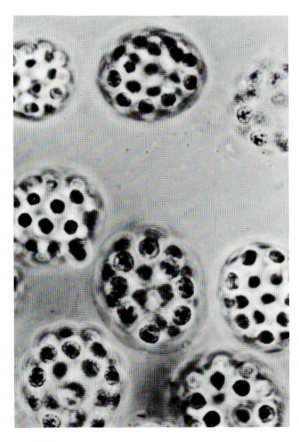

FIG. 8

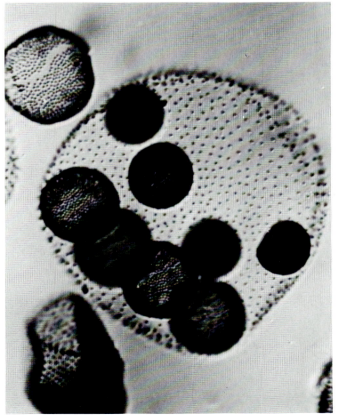

FIG. 9

FIG. 10. *Volvox* life cycle.

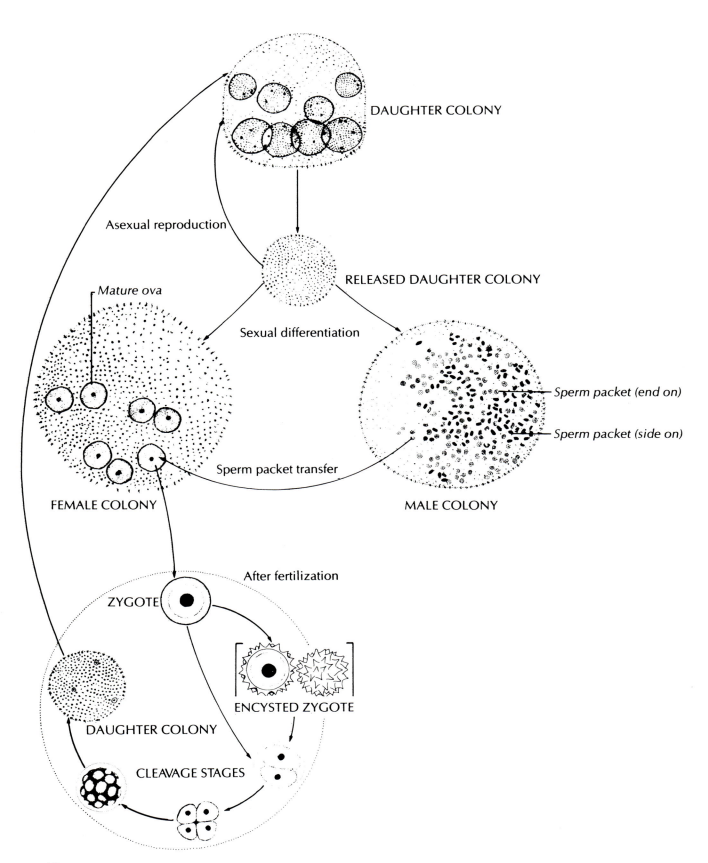

DAUGHTER COLONY

Asexual reproduction

RELEASED DAUGHTER COLONY

Mature ova

Sexual differentiation

Sperm packet (end on)

Sperm packet (side on)

Sperm packet transfer

FEMALE COLONY

MALE COLONY

After fertilization

ZYGOTE

ENCYSTED ZYGOTE

DAUGHTER COLONY

CLEAVAGE STAGES

FIG. 10

FIG. 11. LM, taken using interference optics, of a *Volvox* colony containing female gametes (mature ova) (×330).

FIG. 12. *Volvox* colony containing many male gametes (sperm packets) (×240).

FIG. 13. Following transfer of the sperm packet and fertilization, a zygote is produced (see Fig. 10) that may or may not encyst. The zygote ultimately undergoes cell division (Fig. 13a, ×360 and b, ×350), followed by a unique sort of blastulation that results in a daughter colony (see Fig. 9).

FIG. 11   Mature ova   FIG. 12   Sperm packets

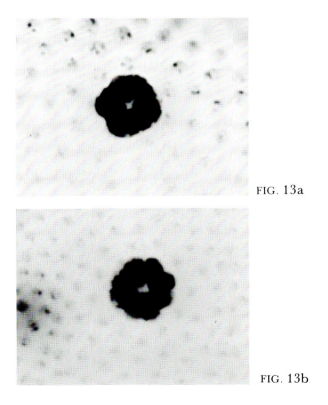

FIG. 13a

FIG. 13b

FIG. 14. SEM of *Ceratium* (×530), an armored dinoflagellate. The pellicle (theca) is composed of a number of highly sculptured plates. The plates lie inside the cell membrane and consist of secreted deposits of cellulose. Dinoflagellates typically have two flagella, a transverse one that partially circles the cell and lies in a groove called the girdle and a longitudinal one that lies in another groove, the sulcus. *(Photograph by E. B. Small, University of Maryland.)*

FIGS. 15 AND 16. SEMs of two *Peridinium* species, also armored dinoflagellates. The plate structure, grooves, and flagella are clearly shown. (Fig. 15, ×1360; Fig. 16, ×750.)

FIG. 17. Dinoflagellate armor is often highly sculptured into wing like structures, as seen in this SEM of *Ornithocercus* (×770). Such sculpturing aids in locomotion and flotation. *Ornithocercus* is fairly common in the marine plankton. This particular specimen was found in a sample of Pacific deep sea snow.

FIG. 18. The plates of some dinoflagellates, for example, *Gyrodinium,* shown here (×1360), are thin and contain little or no cellulose. Often called unarmored or naked dinoflagellates, they have little sculpturing. Several species of naked dinoflagellates, in dense blooms over small areas of the ocean, are responsible for the so-called red tides that are lethal to other marine life.

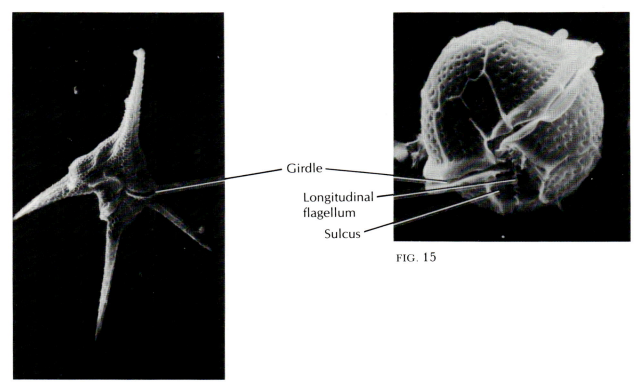

Girdle

Longitudinal
flagellum

Sulcus

FIG. 14

FIG. 15

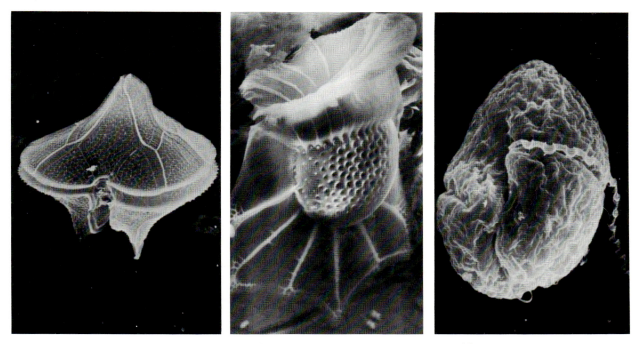

FIG. 16

FIG. 17

FIG. 18

FIG. 22. The various species of amoebas in the Order Amoebida characteristically form one of two types of pseudopods. *Amoeba proteus* (Fig. 22a, ×200) forms blunt, tubular pseudopods, containing both ecto- and endoplasm, called lobopods. Several lobopods often form simultaneously. *(Photograph by E. B. Small, University of Maryland.)* Other amoebas, such as *Saccamoeba* (Fig. 22b, ×400), usually form only a single lobopodial pseudopod. Still other species, such as the shelled amoeba *Arcella discoides,* form very broad, flattened lobopods (Fig. 22c, ×500).

FIGS. 23–26. The shelled amoebas have very distinctive tests (shells). Some species secrete mucopolysaccharide or siliceous tests. The tests of the two species of *Arcella* shown here in SEMs (Figs. 23a, ×680 and 24a, ×1275) are composed of mucopolysaccharide. The fine structure of the test is species-specific (Figs. 23b, ×4730 and 24b, ×9900). (Note the adhering bacteria in Fig. 23b.) The amoeba is attached to the inside of the test and extrudes its pseudopods through an opening in the test.

The tests of other shelled amoebas are made of particles that the cell picks up from the environment. Some of these species such as *Difflugia sp.* (Fig. 25, ×570) use various-sized particles, while other species (Fig. 26, ×770) select only specific sizes of particles. The species shown in Figs. 25 and 26 demonstrate the second type of pseudopod found in this order, the filopod. The filopod is much more slender than the lobopod, may have pointed ends (seen by light microscopy), may be branched, and is filled only with ectoplasm. *(Fig. 25 photograph by E. B. Small, University of Maryland.)*

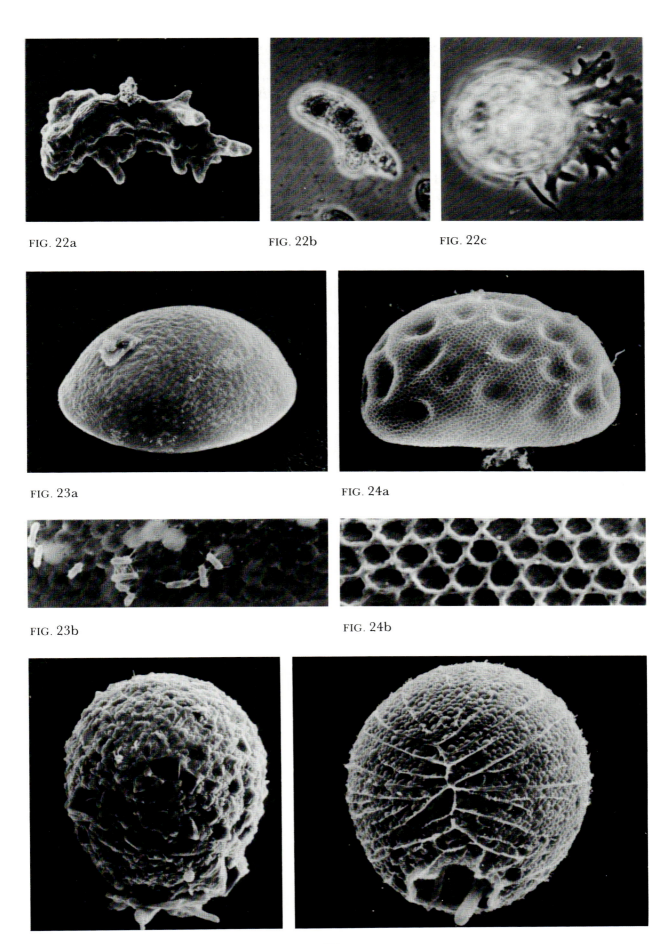

FIG. 22a

FIG. 22b

FIG. 22c

FIG. 23a

FIG. 24a

FIG. 23b

FIG. 24b

FIG. 25

FIG. 26

FIG. 27. The Foraminifera also are protected by a secreted test, but one made of calcium carbonate. The test, as this LM (×50) shows, is often chambered. The forams, both benthic and planktonic species, are very abundant in the marine environment. The planktonic species usually have thinner tests and spines to assist in flotation. The tests of forams are the major constituent of some beach sands as well as of submarine sediments. These calcareous oozes are commonly encountered down to depths of approximately 5,000 meters; below that level the extreme pressures force the calcium carbonate into solution.

FIGS. 28–31. SEMs of foraminiferan tests. The outer surface of the test is covered by a cytoplasmic mantle, which is connected to the internal cytoplasm through a single foramen (aperture). Pseudopods radiate outward from the mantle. The pseudopods are always branched and connected to one another; they are called reticulopods. (Fig. 28, ×80; Fig. 29, ×190; Fig. 30, ×130; Fig. 31, ×200.)

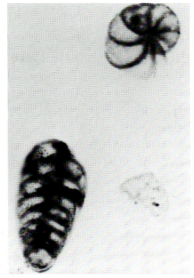

FIG. 27

FIG. 28

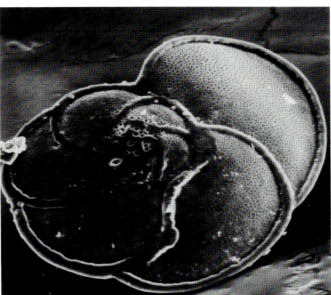

FIG. 29

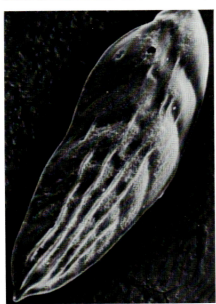

FIG. 30

FIG. 31

FIG. 32. Like the forams, the radiolarians have an internally secreted test, but one made of silica rather than calcium carbonate. The radiolarians are exclusively marine and mostly planktonic. The wide diversity in test morphologies among species is shown in this LM (×120) of a radiolaria "strew." Also like the forams, radiolarian tests often make up substantial portions of oceanic muds, called radiolarian or siliceous oozes. Since silicates are less soluble than carbonates, radiolarian oozes often predominate at depths greater than 5,000 meters.

FIGS. 33–35. SEMs of radiolarian tests. Radiolaria may have filopods or yet a different type of pseudopod, called axopods. An axopod is a very thin pseudopod surrounding a central skeletal core, the axial rod, which is made up of a bundle of microtubules. Some radiolarians have reticulopods (Fig. 33, ×400; Fig. 34, ×410; Fig. 35, ×400). *(Fig. 35 photograph courtesy of E. B. Small, University of Maryland.)*

FIG. 36. The heliozoans may be naked, or be covered externally with foreign particles, or have a siliceous test. All have a large number of axopods radiating outward. This LM (×160) of *Actinosphaerium* shows the axopods clearly. The bulge at the right side of the cell is a contractile vesicle. Heliozoans are primarily found in fresh water and may be planktonic or benthic.

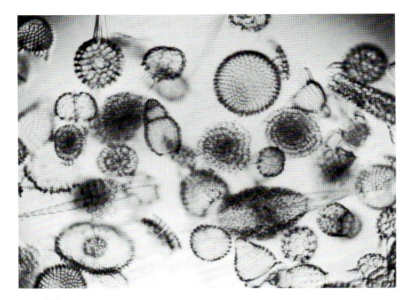

FIG. 32

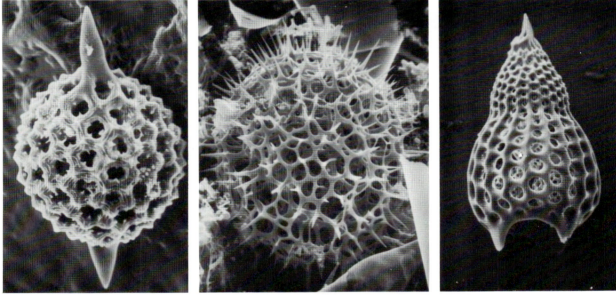

FIG. 33

FIG. 34

FIG. 35

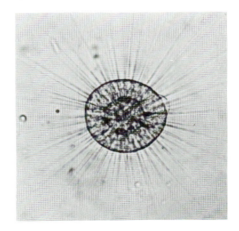

FIG. 36

FIGS. 37 AND 38. LM of *Paramecium,* using phase-contrast optics (Fig. 37, ×180). Both of the fixed-position contractile vacuoles (diagrammed in Fig. 38) are in full diastole, and a food vacuole is forming at the end of the cytopharynx.

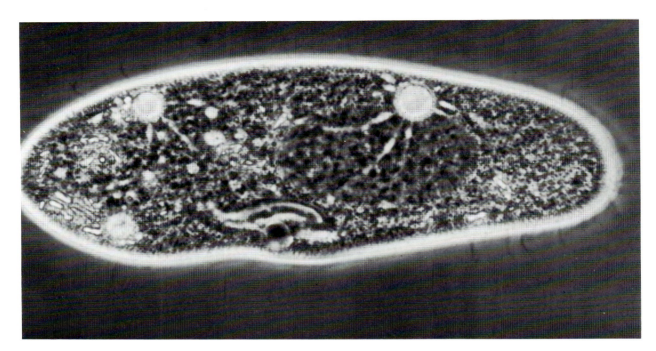

FIG. 37

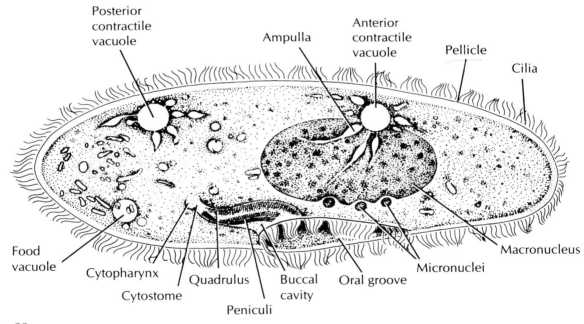

FIG. 38

FIG. 39. SEM (×420) of *Paramecium.* The metachronal wave of ciliary beat has been preserved by the fixative. *(Photograph by E. B. Small, University of Maryland.)*

FIG. 40. Immediately underlying the pellicle in *Paramecium* are structures called trichocysts. Under certain conditions (acid, electrical stimulation), the trichocysts explosively discharge as fibrous threads with pointed tips (Fig. 40a, ×31,500). Discharged trichocysts are visible along the right-hand side of the *Paramecium* in the phase-contrast LM (Fig. 40b, ×525). The shorter hairs on the left of the cell are cilia. This *Paramecium* was exposed to acidified methylene blue. The long, striated trichocyst shafts are obvious in the SEM (Fig. 40c ×23,000). Defense, offense, and anchorage have been suggested as uses for the trichocysts over the years, but the function of these structures in *Paramecium* remains unknown. *(Fig. 40c photograph by E. B. Small, University of Maryland.)*

FIG. 41. The pellicle of ciliates is a very complex structure. It is covered by a surface membrane. The sculpturing is due to an underlying network of kinetodesmal fibers that are associated with the cilia, microtubules, and trichocysts. A nigrosin-stained pellicle (Fig. 41a, ×350) clearly maps the ciliary pits. A close-up SEM of the pellicle (Fig. 41b, ×8600) shows more detail. The knobs in the middle of each pit are broken cilia shafts. Notice that some pits contain more than one cilium.

FIG. 42. LM (×275) showing a pair of *Paramecium* conjugating; the nuclear transformations are well under way.

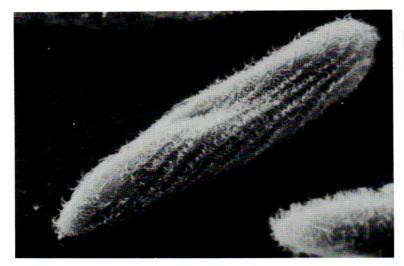

FIG. 39

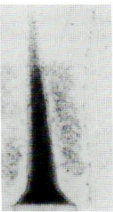

FIG. 40a

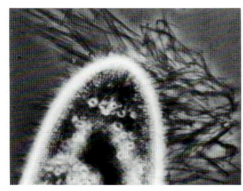

FIG. 40b

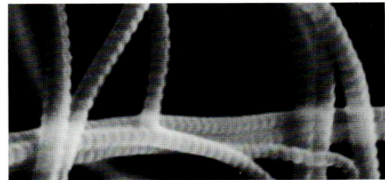

FIG. 40c

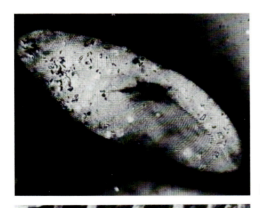

FIG. 41a

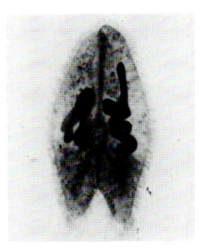

FIG. 42

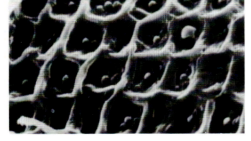

FIG. 41b

FIG. 43. *Didinium* is a carnivorous ciliate. It swims rapidly using two bands of cilia (Fig. 43a, ×720). (The rest of the cell is not ciliated.) *Didinium* eats *Paramecium*. When it comes in contact with a *Paramecium*, a *Didinium* immobilizes its prey with toxic trichocysts called toxocysts. The pointed end of the *Didinium*, called the proboscis (actually an everted cytopharynx; Fig. 43b, ×1360), opens, a complex action involving underlying microtubules. The *Paramecium* is then ingested (Fig. 43c, ×530). Note that the portion of the *Paramecium* about to be ingested has been de-ciliated. Also, the *Paramecium* has fired some trichocysts, but to no avail.

FIG. 44. Ciliates reproduce by transverse fission. These two SEMs show *Didinium* dividing. The initial stages begin with the production of two additional ciliary bands (Fig. 44a, ×700). Fig. 44b (× 650) shows the two daughter cells about to separate. *(All photographs courtesy of E. B. Small, University of Maryland.)*

26

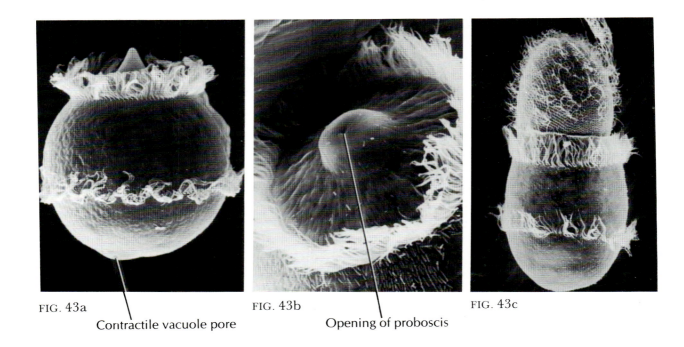

FIG. 43a

Contractile vacuole pore

FIG. 43b

Opening of proboscis

FIG. 43c

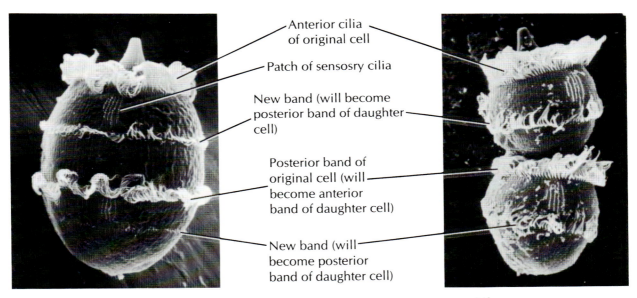

Anterior cilia
of original cell

Patch of sensosry cilia

New band (will become
posterior band of daughter
cell)

Posterior band of
original cell (will
become anterior
band of daughter cell)

New band (will
become posterior
band of daughter cell)

FIG. 44a

FIG. 44b

FIGS. 45 AND 46. Groups of cilia are often fashioned into complex structures having diverse functions. For example, the cilia partially surrounding the oral opening in *Tetrahymena* (Fig. 45, ×2580) form a feeding structure called the undulating membrane. The body ciliature of *Euplotes* (Fig. 46 ×600) is much reduced. Many cilia have grouped together to form a few cirri, which the cell uses to crawl about the substrate. Other cilia have fused to form a membranelle lining the entrance to the buccal cavity.

FIG. 47. LM showing the trumpet-shaped *Stentor* attached to a piece of debris (Fig. 47a, ×180). *Stentor* can also detach, round up and swim off (Fig. 47b, LM using interference optics, ×200). At the oral end of *Stentor* is a complex of ciliated membranelles used in feeding (Fig. 47c, SEM, ×770). *(Photographs in Figs. 45 and 47a by E. B. Small, University of Maryland.)*

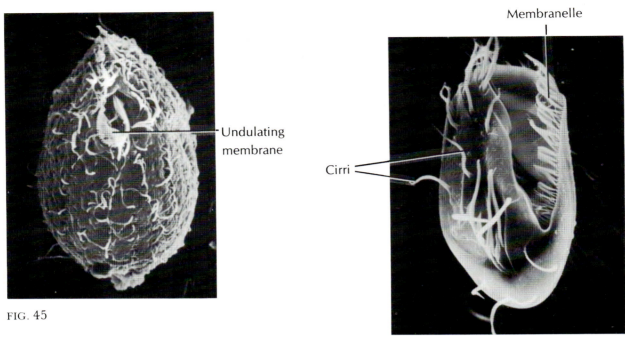

FIG. 45

FIG. 46

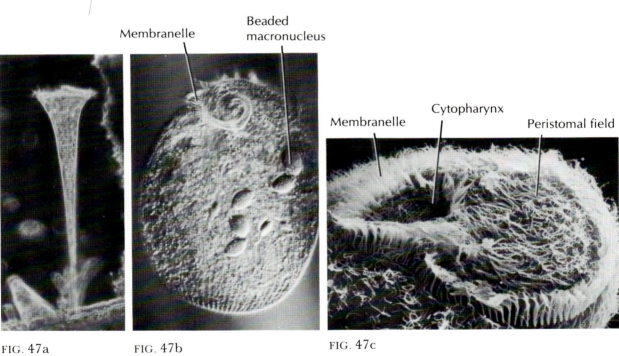

FIG. 47a

FIG. 47b

FIG. 47c

FIGS. 48–51. Most peritrichous ciliates attach to the substrate by stalks that are often highly contractile. For example, *Vorticella* (Fig. 48a, LM, ×136), shown feeding, helically contracts its stalk and its feeding ciliature when disturbed (Fig. 48b, ×680). An organelle called a myoneme, which runs the length of the stalk and has contractile properties, controls the stalk contraction. Other peritrichs form colonies, for example, *Epistylis* (Fig. 49, ×90), *Carchesium* (Fig. 50a, ×200 and b, ×1220), and *Zoothamnium* (Fig. 51a, ×380 and b, ×2000). *(Photographs of Figs. 48, 49 and 50 by E. B. Small, University of Maryland.)*

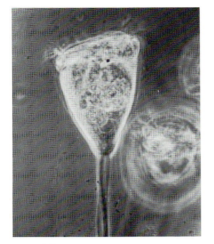

FIG. 48a

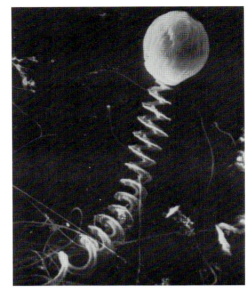

FIG. 48b

FIG. 49   FIG. 50a   FIG. 50b

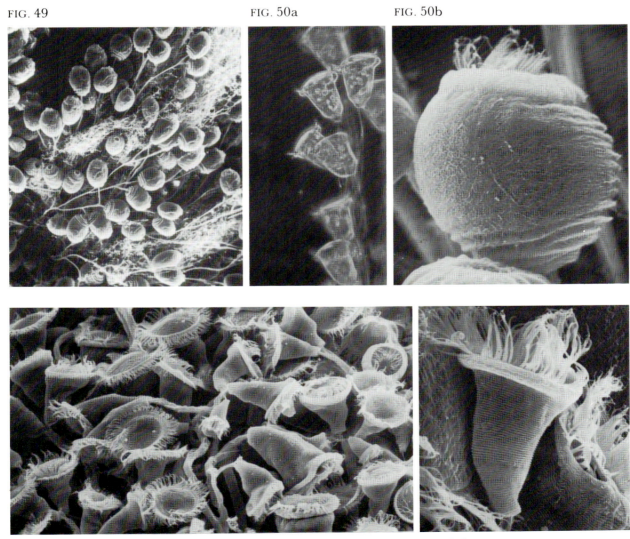

FIG. 51a   FIG. 51b

SECTION II

Taxonomic Summary

KINGDOM ANIMALIA
Phylum Porifera
CLASS CALCAREA
    Order Homocoela
        Leucosolenia
    Order Heterocoela
        Scypha (Grantia)
CLASS HEXACTINELLIDA
        Euplectella
CLASS DEMOSPONGIA
        Microciona
        Spongilla

FIGS. 52 AND 53. *Leucosolenia,* a calcareous sponge, typifies the simplest sponge body type, the asconoid. Its body consists of a simple tube (Fig. 52, ×90) that is covered on the outside by a thin layer of cells, the pinacocytes. This cell layer is clearly seen in close-up in the SEM (Fig. 53, ×430). Monaxon and triaxon calcareous spicules are also obvious. Scattered among the pinacocytes are cells called porocytes, through which water enters the sponge. Water exits the sponge through a single excurrent opening, the osculum (Fig. 52).

FIGS. 54 AND 55. The inside of the body tube of an asconoid sponge is lined with cells called collar cells or choanocytes. (The choanocyte layer is visible just inside the oscula in Fig. 52.) These cells are all flagellated, and the beating flagella draw water through the porocytes into the center of the asconoid tube, the spongocoel. The flagellum of each choanocyte is surrounded by a collar of microvilli (Fig. 54, ×1700). The collar filters food particles from the water current created by the flagellum.

The choanocyte is very similar in morphology (Fig. 55) to the individual cells of a unique group of colonial flagellate protozoa, the choanoflagellates. This similarity is seen by some as evidence of a close evolutionary link between the sponges and the colonial flagellates. Others view the similarity as an indication of convergent evolution.

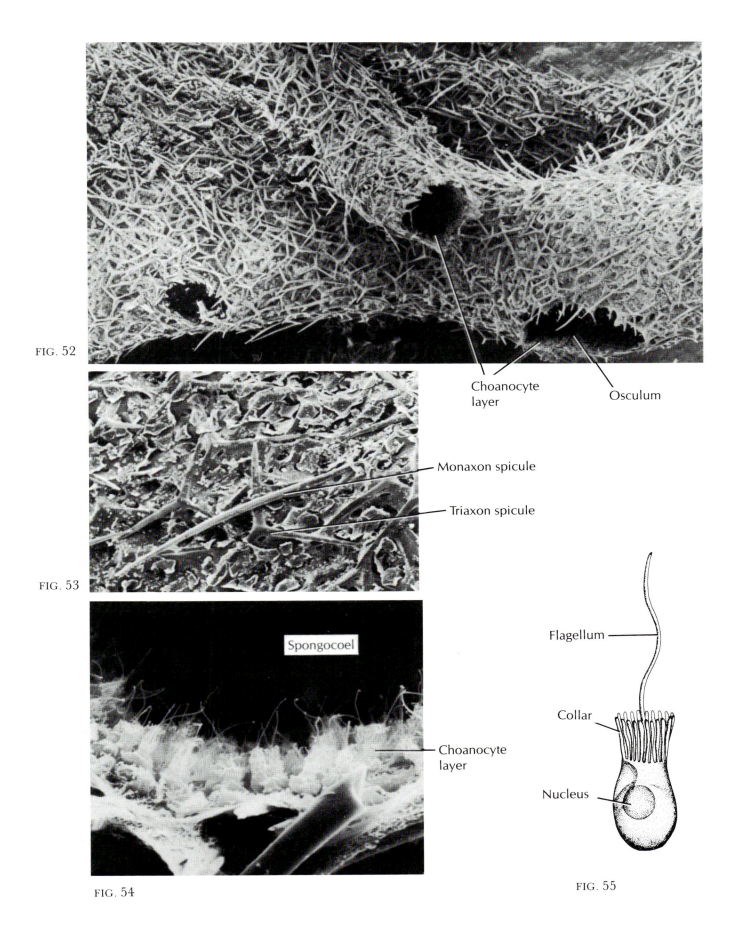

FIG. 52

Choanocyte layer

Osculum

FIG. 53

Monaxon spicule

Triaxon spicule

Spongocoel

Choanocyte layer

FIG. 54

Flagellum

Collar

Nucleus

FIG. 55

FIGS. 56 AND 57. Another calcareous sponge, *Scypha,* represents a slightly more complex sponge body type, the syconoid grade. A longitudinal section of *Scypha* is shown in the LM (Fig. 56, ×30). While the syconoid sponge is still basically tubular overall, the choanocyte layer has become infolded and an acellular body wall matrix, the mesohyl, fills the spaces between the flagellated chambers (Fig. 57). Spicules are imbedded in the gelatinous mesohyl and amoeboid cells are also present in this layer. Syconoid sponges do not have porocytes; the incurrent openings, called ostia, are invaginations of the body surface.

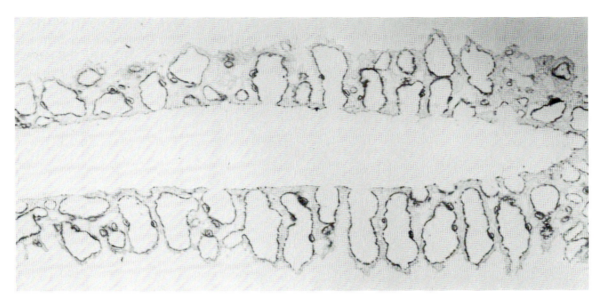

FIG. 56

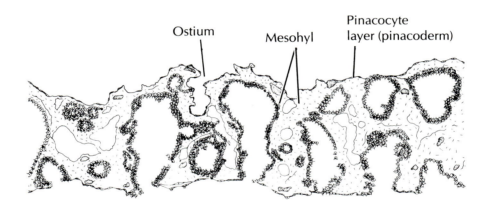

Ostium   Mesohyl   Pinacocyte layer (pinacoderm)

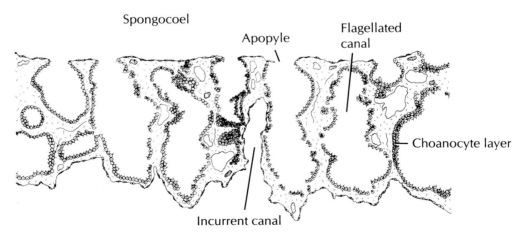

Spongocoel   Apopyle   Flagellated canal

Incurrent canal   Choanocyte layer

FIG. 57

FIGS. 58 AND 59. *Scypha* in cross section (Fig. 58, LM, ×86). As in the asconoid sponges, the flagella of the choanocytes draw water into the spongocoel. The openings into the flagellated canals from the incurrent canals are called prosopyles (Fig. 59). The flagellated canals open into the spongocoel through the apopyles.

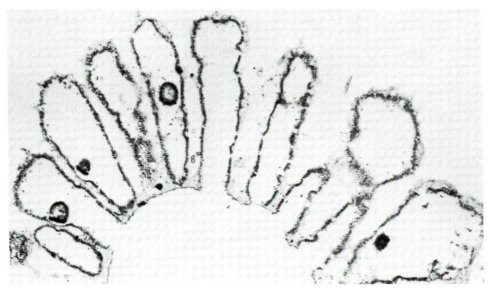

FIG. 58

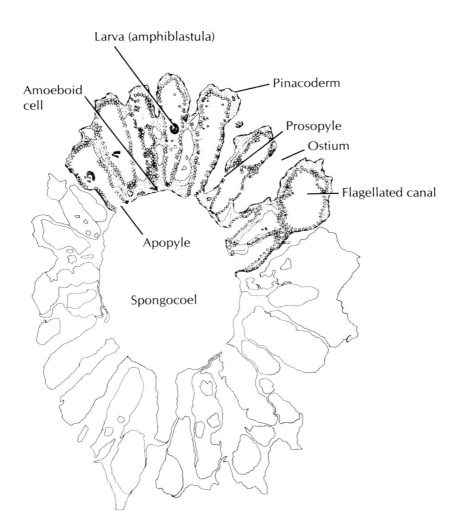

FIG. 59

FIG. 60. *Euplectella* is a hexactinellid sponge, or glass sponge. Its spicules, six-rayed and made of silica, form a skeleton. Many of the hexactinellid sponges are found in the deep sea. Their body type is syconoid grade.

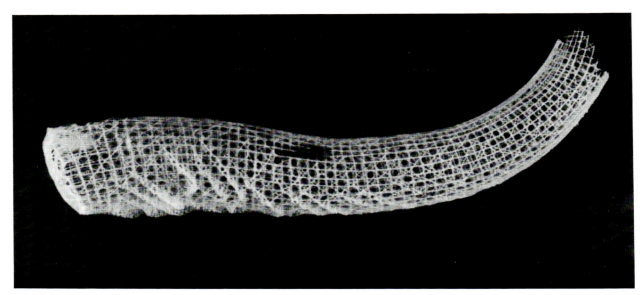

FIG. 60

FIGS. 61–63. SEMs (Fig. 61, ×490; Fig. 62, ×1490; Fig. 63, ×3650) of cryofractured *Microciona* showing the complexity of the third sponge body type, the leuconoid. The body wall is a much-thickened mesohyl cortex. The choanocytes are located in flagellated chambers deep within the mesohyl. A highly branched series of incurrent and excurrent canals brings the water through the cortex. The inhalant openings of these canals are called dermal pores, but they are not porocytes. The excurrent canals usually anastomose to form a few centrally located oscula.

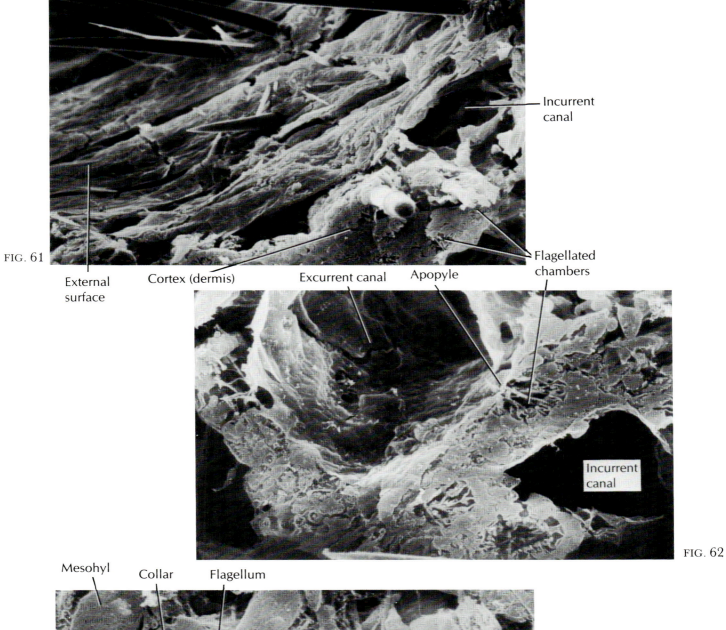

Incurrent
canal

FIG. 61

External
surface

Cortex (dermis)

Excurrent canal

Apopyle

Flagellated
chambers

Incurrent
canal

FIG. 62

Mesohyl   Collar   Flagellum

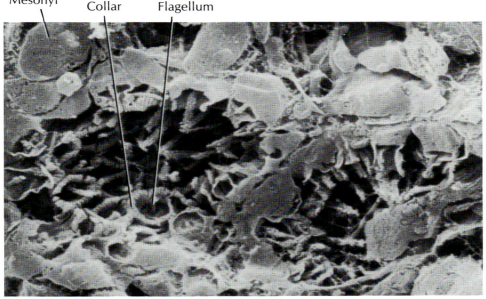

FIG. 63

FIGS. 64–67. Sponges can regenerate and all have some capacity to reaggregate to form a complete sponge following dissociation into single cells. These SEMs show the reaggregation of *Microciona prolifera*.

A piece of *Microciona* was squeezed through several layers of cheese cloth, which dissociated the body wall into a suspension of single cells (Fig. 64, ×460). The cells immediately became amoeboid, creeping around until they came in contact with other cells, stuck together, and gradually reformed the sponge. Though accounts in the literature indicate that the choanocytes lose their collars and flagella during the amoeboid phase, on careful inspection, amoeboid choanocytes are evident in all stages of the reaggregation process (Fig. 65, ×4,900). Six hours after dissociation, small clumps of reaggregated cells occurred throughout the cell suspension (Fig. 66, ×280). These clumps attracted more and more cells, continually growing in size (Fig. 67, after 12 hours, ×50) until a new sponge was produced. *Microciona* seems to be the only species that can reaggregate completely; other species reaggregate to a lesser extent. The phenomenon is also species-specific. Cells from one species will not usually reaggregate with cells from another species.

FIG. 64

FIG. 65

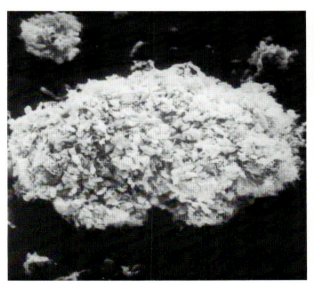

FIG. 66

FIG. 67

FIGS. 68 AND 69. Sponges can reproduce sexually or asexually. They reproduce sexually by shedding sperm into the surrounding seawater. Sperm cells that happen to enter an adjacent sponge fertilize egg cells in the flagellated chambers. Cleavages occur as the embryo develops, resulting in a parenchymula larva in most species. In some of the calcareous sponges, a flagellated amphiblastula larva is produced. Fig. 68 ($\times$400) is a LM of a *Scypha* amphiblastula inside a flagellated canal. The amphiblastula leaves the parent sponge and usually attaches to the substrate. It then undergoes a form of gastrulation that relocates the flagellated cells to the inside of the larva. The flagellated cells become the initial choanocytes. The macromeres (Fig. 69) are destined to become the first pinacocytes.

FIG. 70. Sponges reproduce asexually by budding. In addition, all freshwater sponges and some marine forms produce asexual reproductive structures called gemmules. A gemmule of the freshwater sponge *Spongilla* is shown here (LM, $\times$64). The gemmule is a ball of food-filled amoebocytes surrounded by a tough secreted membrane and a wall of spicules. The amoebocytes leave the gemmule through the micropyle during germination. Gemmules in freshwater sponges are usually produced before winter; they germinate in the spring.

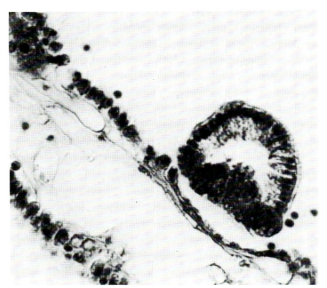

FIG. 68

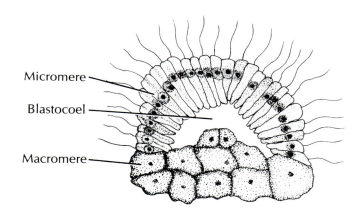

Micromere

Blastocoel

Macromere

FIG. 69

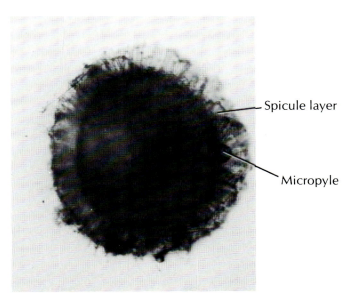

Spicule layer

Micropyle

FIG. 70

SECTION III
Taxonomic Summary

Phylum Cnidaria
  CLASS HYDROZOA
    Order Hydroida
      Suborder Anthomedusae
        Hydra
        Tubularia
        Pennaria
        Hydractinia
      Suborder Leptomedusae
        Obelia
        Campanularia
        Schizotricha
    Order Siphonophora
        Physalia
  CLASS SCYPHOZOA
    Order Semaeostomeae
        Aurelia
        Chrysaora
        Cyanea
    Order Rhizostomae
        Cassiopeia
  CLASS ANTHOZOA
   SUBCLASS ALCYONARIA (OCTOCORALLIA)
    Order Gorgonacea
        Leptogorgia
    Order Pennatulacea
        Renilla
   SUBCLASS ZOANTHARIA
    Order Actinaria
        Metridium
        Diadumene
    Order Scleractinia (Madreporaria)
        Astrangia

FIG. 71. The life cycles of Cnidarian species are characterized by the presence of two types of adult morphology: a sessile form, the polyp, and a free-swimming form, the medusa. In the class Hydrozoa, the polyp form is usually colonial, with the colony composed of at least two types of polyps, both of which are shown here in *Obelia*.

The main function of the gastrozooid polyp is feeding. The tentacles contain specialized cells, the cnidoblasts, that fire nematocysts of various kinds (see Figs. 130–133); the nematocysts immobilize the prey either mechanically or toxically. The gastrozooid polyps have mouths to ingest the prey, and their body cavity—the gastrovascular cavity or coelenteron, which serves as both digestive system and hydrostatic skeleton—is continuous throughout the colony. The gastrozooid also has a role in colony defense.

The function of the other polyp, the gonozooid, is reproduction. Often, the gonozooid is highly specialized for this task: in many species it lacks tentacles and a mouth, and must rely on the gastrozooids for food and defense. Typically, the gonozooid polyp produces medusae by budding, as in *Obelia*. Once a medusa buds from the gonozooid, it produces gametes. Fertilization occurs and results in a ciliated larva, the planula, that attaches to the substrate and develops into a colony.

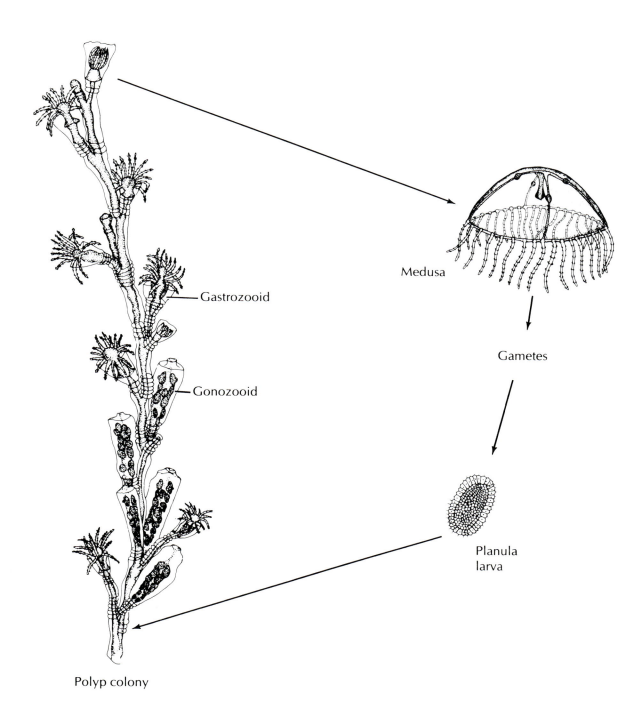

Medusa

Gastrozooid

Gonozooid

Gametes

Planula
larva

Polyp colony

FIG. 71

FIGS. 72 AND 73. Many hydrozoan species do not have free-swimming medusa stages, although some vestige of the medusa usually remains in the life cycle (freshwater hydras are an exception). While *Obelia* (Fig. 72, LM, ×30) does have free-swimming medusae, *Campanularia* (Fig. 73, LM, ×15) does not. Medusoid remnants are the sessile gonophores within the gonozooid polyps. Both *Obelia* and *Campanularia* are in the suborder Leptomedusae, whose members characteristically have hydranths surrounded by hydrothecas, which are extensions of the skeletal perisarc.

FIGS. 74 AND 75. In *Tubularia*, in the suborder Anthomedusae (in which the perisarc, if present, does not cover the hydranth) the medusoid stage is further degenerated. There are no gonozooids in *Tubularia*. The gonophores, shown here in SEMs (Fig. 74, ×20; Fig. 75, ×140), develop on the hydranth between the rings of proximal and distal tentacles. Not only is there no free-living medusa, but the planula develops inside the gonophore and an actinula larva (see Fig. 80) is ultimately released.

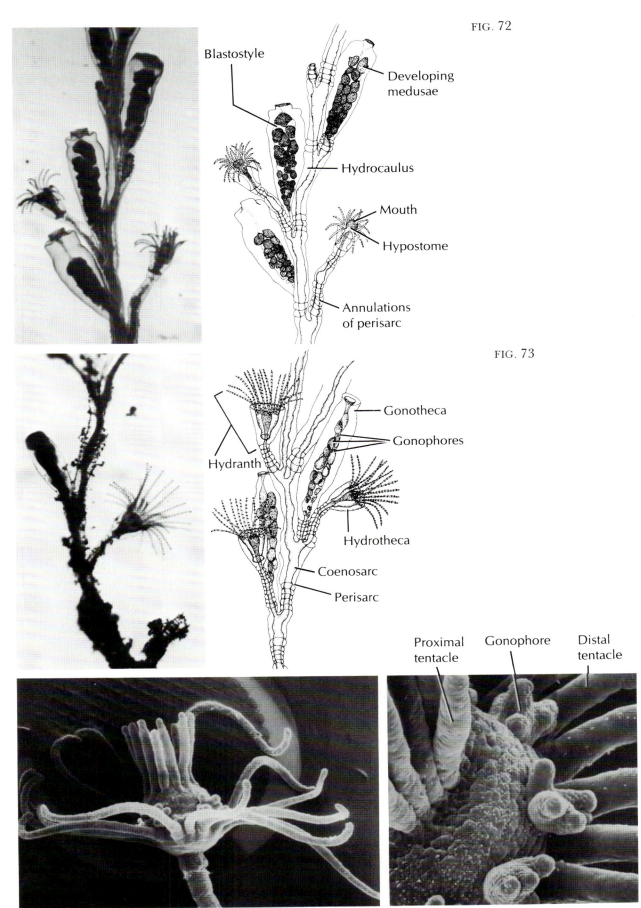

FIG. 72

Blastostyle

Developing
medusae

Hydrocaulus

Mouth

Hypostome

Annulations
of perisarc

FIG. 73

Gonotheca

Gonophores

Hydranth

Hydrotheca

Coenosarc

Perisarc

Proximal
tentacle

Gonophore

Distal
tentacle

FIG. 74

FIG. 75

FIGS. 76–79. SEMs of the hypostome and mouth of an *Obelia* gastrozooid (Fig. 76, ×340). The hypostome is surrounded by tentacles, which have annular rings of cnidoblasts along their length. The cnidocils, which have a flagellumlike ultrastructure, project out of each cnidoblast (Fig. 78 ×1900) and are involved in recognizing and transducing the stimuli that cause the nematocysts to fire. The distribution of the cnidoblasts along the tentacles varies from species to species. For example, the cnidoblasts are staggered along the tentacles of *Campanularia* gastrozooids (Fig. 79, ×720).

Various organisms live attached to hydrozoan colonies. Bacteria, diatoms, and stalked protozoa are attached to the hydrotheca of the *Obelia* polyp in Fig. 77 (×330).

FIG. 80. *Tubularia* actinula larvae (LM, ×30).

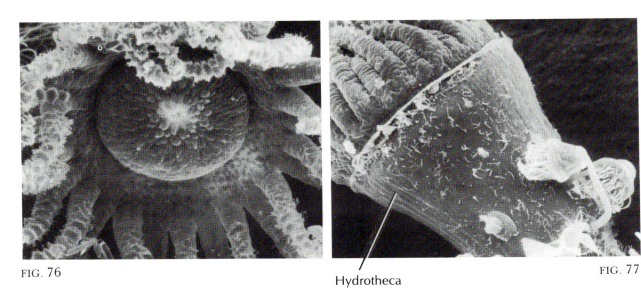

FIG. 76

FIG. 77

Hydrotheca

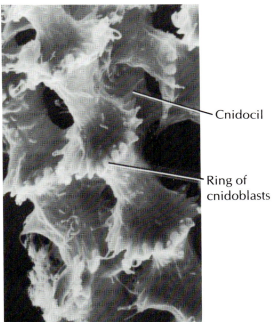

Cnidocil

Ring of
cnidoblasts

FIG. 78

FIG. 79

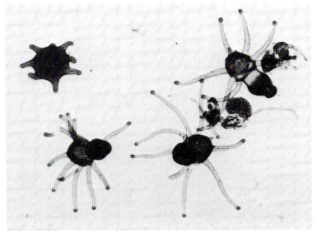

FIG. 80

FIG. 81–83. *Pennaria*, also an anthomedusan species, demonstrates another alternative in medusa production. As in *Tubularia*, there are no gonozooids *per se* in the *Pennaria* colony (Fig. 81). The gonophores form on the hydranths, sometimes more than one on a single hydranth (Fig. 82, LM; Fig. 83, SEM, ×70). Each gonophore will become a single medusa and eventually buds off the polyp. The free-swimming medusae produced by *Pennaria* are partially degenerate—the tentacles never become more than bumps—and are relatively short-lived.

FIG. 81

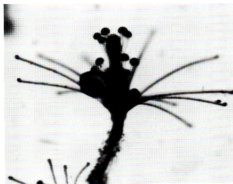

FIG. 82a

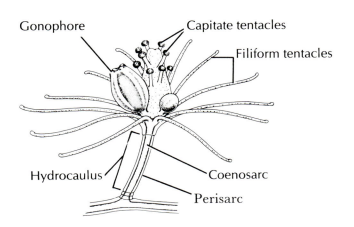

Gonophore   Capitate tentacles

Filiform tentacles

Hydrocaulus

Coenosarc

Perisarc

FIG. 82b

FIG. 83

FIG. 84.   SEM (×260) of mouth and hypostome of a *Pennaria* polyp. The batteries of nematocysts on the ends of the knobbed capitate tentacles are evident.

FIG. 85.   In addition to attached protists, many metazoa often can be found on hydrozoan colonies. Some of them eat the hydrozoan polyps. Others, such as the water bear (phylum Tardigrada), shown here (×460) on the end of a capitate tentacle, are probably scavengers.

FIGS. 86 AND 87.   Some colonial hydrozoa have other polyps in addition to the feeding and reproductive types. These SEMs (Fig. 86, ×200; Fig. 87, ×660) show *Schizotricha*, a hydrozoan with small polyps called nematophores at intervals along the hydrocaulus and at the base of the hydrothecae. The nematophores function in prey capture; their tentacles may contain adhesive cells as well as cnidoblasts.

FIG. 84

FIG. 85

FIG. 86

FIG. 87

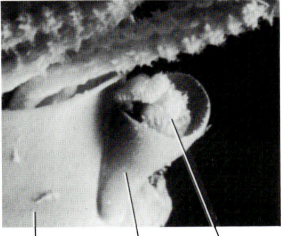

Hydrotheca   Nematotheca   Nematophore
of
gastrozooid

FIGS. 88–90. *Hydractinia,* an anthomedusan hydrozoan, forms a colony consisting of several types of polyps (Fig. 88, LM, ×20). Along with gastrozooids, *Hydractinia* has a polyp type, the dactylozooid, that has short, capitate tentacles, no mouth, is extensible, and functions in defense and food capture (Fig. 89, SEM, ×100). The dactylozooids rely on the rest of the colony for their food. The gonozoids of *Hydractinia* usually bear sessile gonophores (Fig. 90, ×220). Both male and female gonozoids have been described. The *Hydractinia* colony is studded with projections of the perisarc (also called periderm) called epidermal spines.

The habitat of *Hydractinia* is unique: it encrusts the shells of hermit crabs. An entire colony attaches to a hermit crab shell with many layers of periderm. The polyps project through the perisarc layer and are not covered by it.

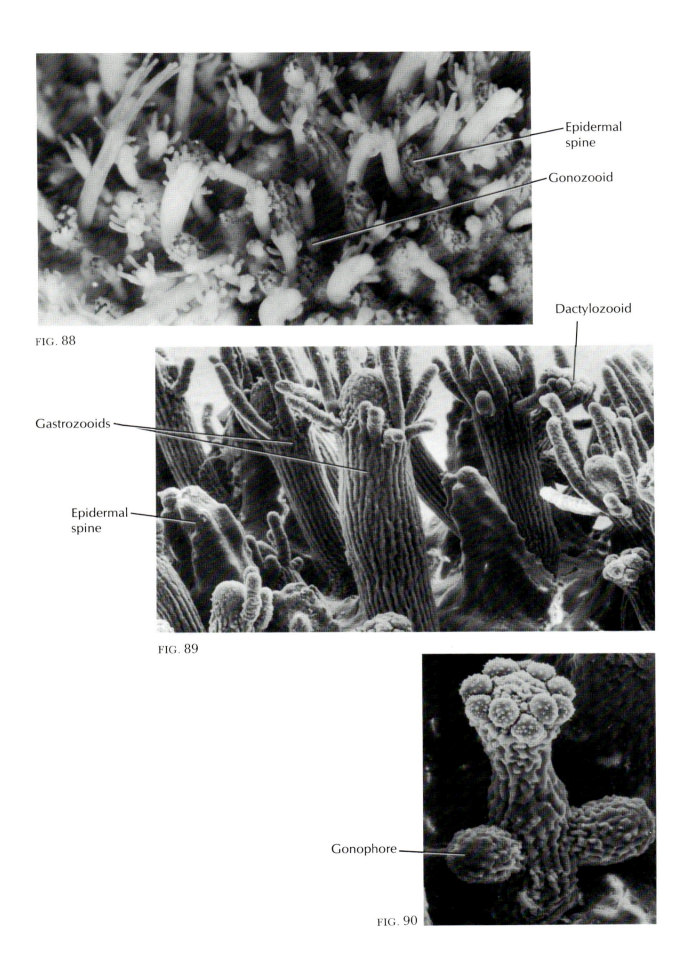

Epidermal spine

Gonozooid

FIG. 88

Dactylozooid

Gastrozooids

Epidermal spine

FIG. 89

Gonophore

FIG. 90

FIGS. 91 AND 92. One of the most spectacular examples of polyp polymorphism is *Physalia,* the Portuguese man-of-war (suborder Siphonophora). *Physalia* is a floating colony (Fig. 91). The float, called the pneumatophore, seems to be a modified medusa, although it has been called a modified polyp by some. It is a sac filled with gas secreted by a gas gland. (Older reports indicate that the composition of the gas inside the *Physalia* float is very similar to that of air. In other siphonophores, the gas is carbon monoxide.)

The dactylozooids (Fig. 92) have a single tentacle and no mouth, and are extremely extensible. Coiled around the length of the dactylozooid are batteries of nematocysts. The gastrozooids have a mouth and usually a single contractile tentacle. They are much shorter than the dactylozooids and function only in food ingestion. The gonozooids are usually branched (occasionally called gonodendra) and clusters of the sessile gonophores (called gonopalpons) are found on the tips of the branches. All of the individual polyps of a *Physalia* colony bud from the coenosarc at the base of the pneumatophore.

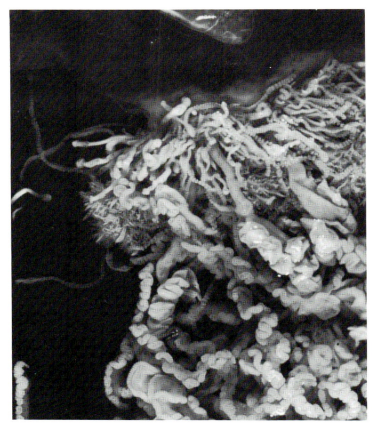

FIG. 91

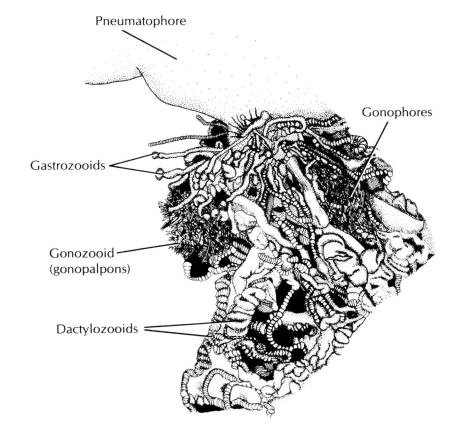

Pneumatophore

Gonophores

Gastrozooids

Gonozooid
(gonopalpons)

Dactylozooids

FIG. 92

FIGS. 93 AND 94. Hydrozoan medusae are fairly small and usually short-lived, and they have a fairly standard body plan. However, the morphology of one of the more familiar hydromedusae, *Obelia* (Fig. 93, LM, ×140; Fig. 94, diagram), does not conform entirely to the standard. Most hydrozoan medusae have a velum, a ring of tissue around the outer edge of the subumbrellar surface, but *Obelia* medusae lack this diagnostic feature.

FIGS. 95 AND 96. LM and diagram of a more typically shaped, although unidentified, hydromedusa. This specimen has a prominent velum and the tentacles are grouped into four bunches at the bottom of each radial canal. Batteries of nematocysts give a dotted appearance to the tentacles in the LM.

FIG. 97. Hydromedusae are extremely abundant in the marine plankton at certain times of the year. This LM shows medusae taken in a plankton tow off Cape Cod during the summer. The taxonomy of hydromedusae is not yet well defined, and in some cases a medusa has its own species designation because its polyp source is unknown.

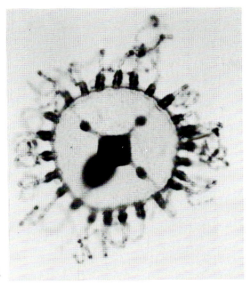

FIG. 93

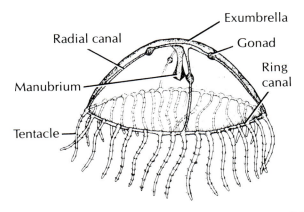

FIG. 94

Radial canal
Exumbrella
Manubrium
Gonad
Ring canal
Tentacle

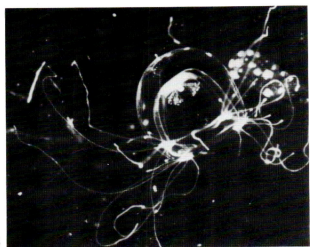

FIG. 95

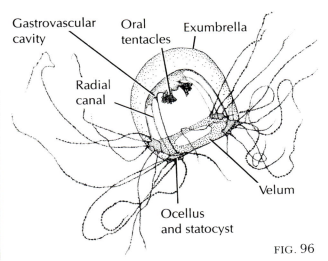

Gastrovascular cavity
Oral tentacles
Exumbrella
Radial canal
Velum
Ocellus and statocyst

FIG. 96

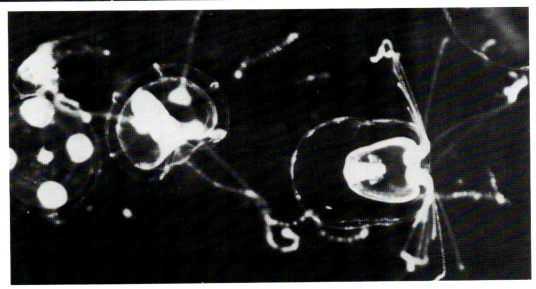

FIG. 97

FIG. 98. The commonly studied, freshwater *Hydra* has no vestige of a medusa stage in its life cycle. *Hydra* usually reproduces by budding (see Fig. 99). Under certain environmental conditions (for example, the onset of winter), special cells called interstitial cells aggregate to form ovaries or testes that, in turn, produce germ cells. An egg remains attached to the female until it is fertilized, after which a chitinous covering is produced. The embryo then drops off the female and remains dormant until conditions are favorable for the hatching of the young hydra.

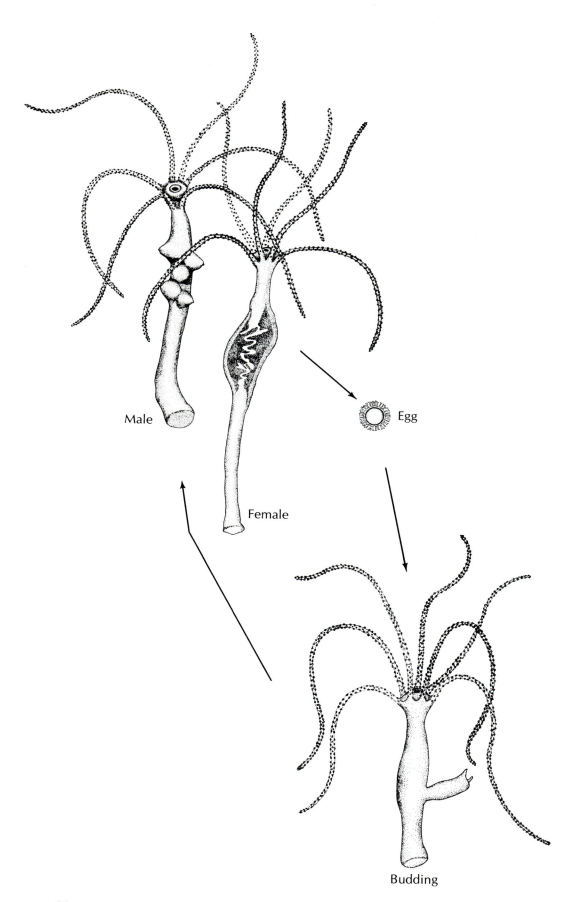

Male

Female

Egg

Budding

FIG. 98

FIG. 99. LM (×22) of a budding *Hydra*. The thin body wall and the gastrovascular cavity, which is continuous with the bud and the tentacles, are obvious. Unlike the marine hydrozoa, *Hydra* is not colonial. It attaches to the substrate by its aboral pedal disc.

FIGS. 100 AND 101. The body wall of cnidarians consists of two tissue layers, an outer ectoderm and an inner gastrodermis, separated by a thin, noncellular mesogloea. No mesoderm is ever present at any stage. These LMs show a cross section (Fig. 100, ×135) and a longitudinal section (Fig. 101, ×175) through the body column of *Hydra*.

The epidermis is composed primarily of epitheliomuscular cells interspersed with sensory cells and interstial cells. Cnidoblasts are present, especially in the tentacles. A developing ovary (Fig. 100) lies between the mesogloea and the epidermis. The contractile elements of the epitheliomuscular cells, the myonemes, constitute a longitudinal contractile layer in the body wall. The gastrodermis is composed of nutritive muscle cells that are flagellated and also contain contractile elements. In the gastrodermis, the myonemes are arranged circularly in the body wall. The nutritive muscle cells are usually filled with food vacuoles. The other major cell type in the gastrodermis is the gland cell, which secretes digestive enzymes into the gastrovascular cavity. The cnidarian nervous system typically consists of a subepidermal and subgastrodermal nerve net (see Fig. 134) composed of bipolar or multipolar neurons, but the *Hydra* nerve net is associated only with the epidermis (although occasional investigators have found evidence of some neurons in the gastrodermis).

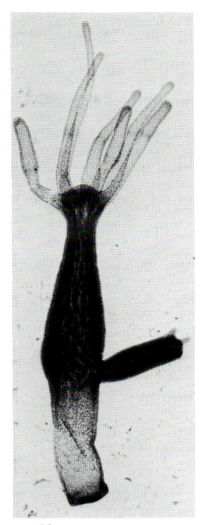

FIG. 99

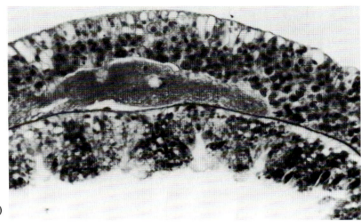

FIG. 100

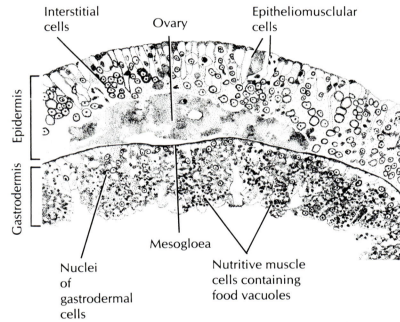

Interstitial cells

Ovary

Epitheliomusclular cells

Epidermis

Gastrodermis

Nuclei of gastrodermal cells

Mesogloea

Nutritive muscle cells containing food vacuoles

Gastrovascular cavity

Mesogloea

Epidermis

Gastrodermis

FIG. 101

FIG. 102. SEM (×35) of a *Hydra* that was relaxed with a dilute Chloretone solution before fixation.

FIG. 103. SEM (×220) of hypostome and mouth of *Hydra*.

FIG. 104. SEM (×280) of *Hydra* tentacle. The nematocyst batteries circle the tentacle at intervals. Note the cnidocils protruding from the tops of the cnidoblasts.

FIG. 105. SEM (×810) showing the cell boundaries of the epidermal cells (epitheliomuscular cells for the most part) at the outer surface of the *Hydra* body column. The small projections that appear to arise from tiny pits may be the flagellar sensors of the epidermal sensory cells.

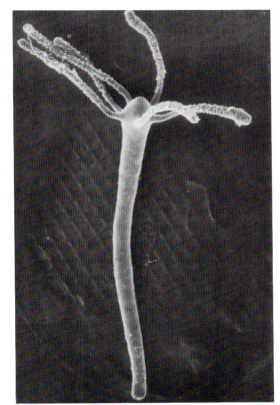

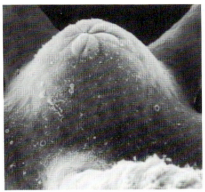

FIG. 103

FIG. 102

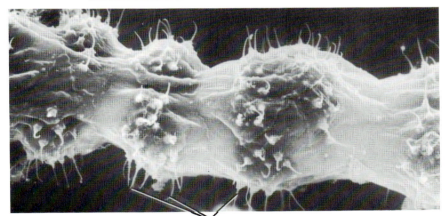

FIG. 104   Cnidocils

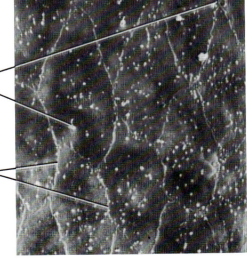

Tips of sensory cells

Cell boundaries

FIG. 105

FIG. 106. The life cycle of *Aurelia* is typical of the scyphozoan medusae. Scyphozoa also have alternate generations, but unlike the Hydrozoa, the medusa is the prominant form. The scyphozoan medusa is usually quite large relative to the polyp and some are long-lived. The medusa sheds gametes into the seawater. After fertilization, a planula larva forms (which temporarily attaches to the oral arms of the adult medusa in *Aurelia* and a few other species—see Fig. 111) that settles and forms a polyp, the scyphistoma. The polyp stage in the Scyphozoa is not usually colonial.

Individual scyphistoma can produce other polyps by budding. Periodically, they also undergo a process called strobilation in which young medusae, called ephyrae, are produced by budding. Many developing ephyrae are produced simultaneously by the scyphistoma, which is called a strobila during this process. After the ephyrae have budded, the polyp resumes its sessile existence, living for years in some species. The ephyrae gradually mature into adult medusae.

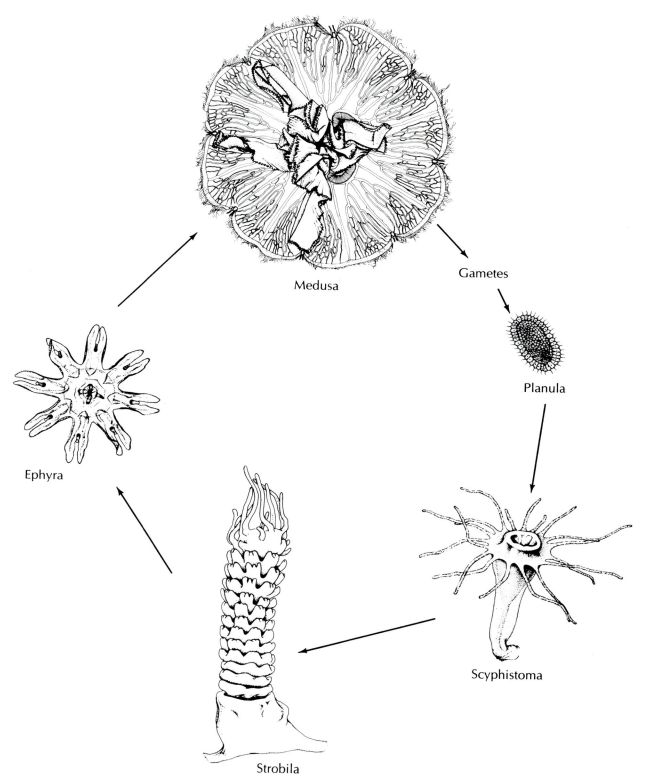

Medusa

Gametes

Planula

Scyphistoma

Strobila

Ephyra

FIG. 106

FIG. 107. Photograph (×40) of scyphistoma of *Aurelia.*

FIG. 108. LM (×25) of a strobila of *Aurelia.* The tentacles at the top of the stack of developing ephyrae will eventually be resorbed and a new set produced under the lowest ephyra. At this stage, a common gastrovascular cavity is present.

FIGS. 109 and 110. LMs (Fig. 109, ×25; Fig. 110, ×73) of an *Aurelia* ephyra. The pulsating contractions of the lappets enable the ephyra to swim. These and other muscular events are initiated and integrated by the nerve net (see Fig. 134) and the marginal ganglia at the notch in each lappet. The complex of each ganglion together with a statocyst and, in some species, a pigment-spot ocellus is collectively called a rhopalium. The rhopalia of the ephyra persist in the adult medusa.

FIG. 111. LM (×80) of planula larvae on the oral arms of *Cyanea.*

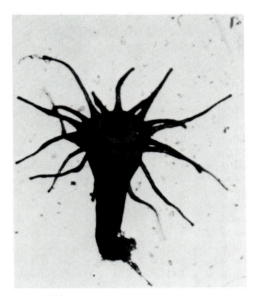

FIG. 107

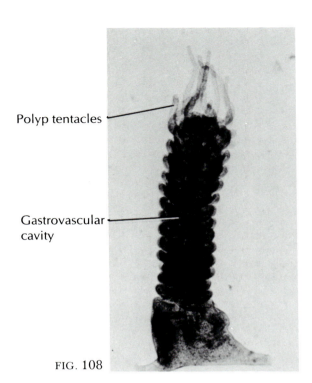

Polyp tentacles

Gastrovascular
cavity

FIG. 108

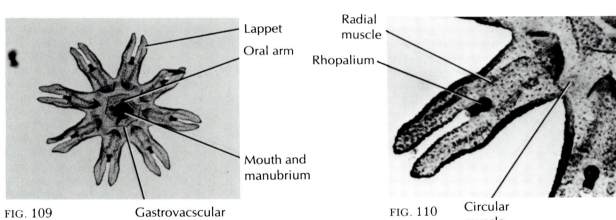

Lappet

Oral arm

Mouth and
manubrium

FIG. 109

Gastrovacscular
cavity

Radial
muscle

Rhopalium

FIG. 110

Circular
muscle

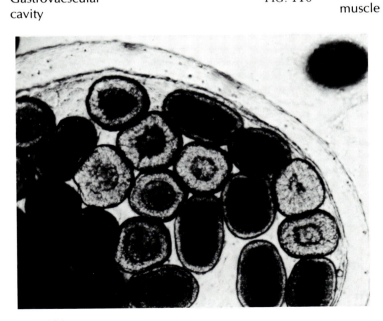

FIG. 111

FIG. 112. Dissection of an *Aurelia* medusa, pinned aboral side down. Although this medusa is fairly common in inshore waters and is also often studied in the classroom, its anatomy differs in some respects from that of the typical scyphozoan jellyfish. In many of the other species, the bell is less flattened and the tentacles along the margin of the bell and the oral arms are longer. The scalloped margin of the *Aurelia* bell is characteristic of scyphozoan medusae.

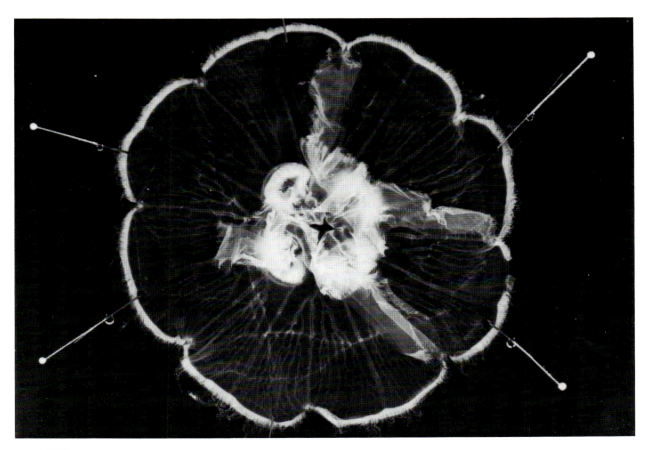

FIG. 112

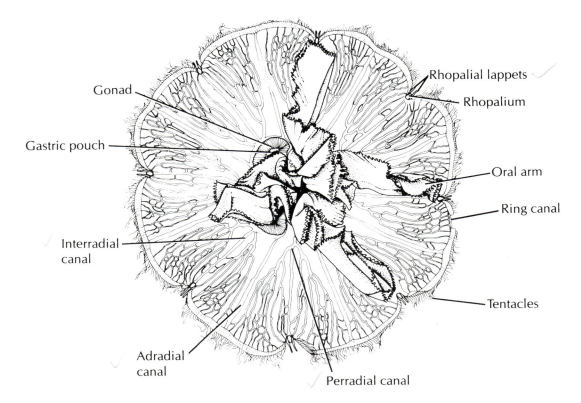

Gonad

Gastric pouch

Interradial
canal

Adradial
canal

Perradial canal

Rhopalial lappets

Rhopalium

Oral arm

Ring canal

Tentacles

FIG. 113. The medusa of *Chrysaora* is more typically constructed than that of *Aurelia*. The marginal tentacles, which contain the nematocysts, are fewer in number and much longer than those of *Aurelia;* they are also highly contractile. The corners of the manubrium are drawn out into long, frilly oral arms that extend farther below the bell than do those of *Aurelia*. Although its anatomy is typically scyphozoan, *Chrysaora* is atypically hermaphroditic.

FIG. 114. Close-up photograph of the subumbrellar surface of *Chrysaora*. The tentacles are contracted.

FIG. 115. The swimming muscle of the scyphozoan medusa is a broad band of circular epidermal muscle called the coronal muscle, shown here in *Cyanea*. It is located on the subumbrellar side of the bell and is partitioned into segments by mesogloea. The tentacles in *Cyanea* occur in clusters attached to the subumbrellar surface. The white dots on the tentacles are nematocysts.

FIG. 116. Close-up of a *Cyanea* rhopalium. The statocyst is a balance-sensing organ. In addition to the structures labeled, the rhopalium also contains a light-sensitive ocellus and the so-called sensory pits.

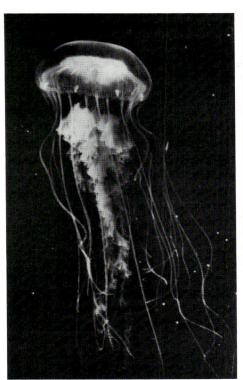

FIG. 113

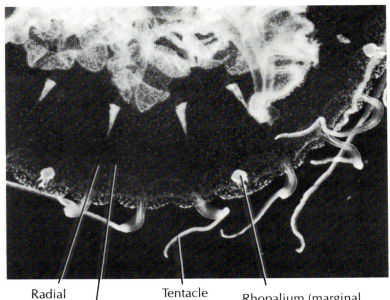

FIG. 114

Radial canal   Subumbrellar surface   Tentacle   Rhopalium (marginal body)

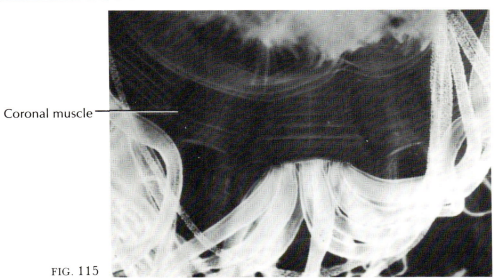

Coronal muscle

FIG. 115

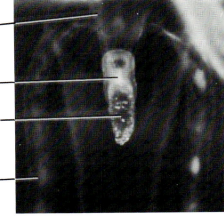

Nerve net

Ganglion

Statocyst

Rhopalial tentacle

FIG. 116

FIGS. 119 AND 120. In the class Anthozoa, there is no medusa stage in the life cycle; only the polyp exists. Both solitary and colonial species occur. Fig. 119 shows a cross-sectional dissection of *Metridium,* an actinarian sea anemone. Fig. 120 (×30) shows a longitudinal section through the column of a smaller specimen of *Metridium.*

The anthozoan polyp anatomy is more complex than that of the hydrozoan polyps. The mouth is located in the center of the oral disc and opens into a central pharynx that is suspended by septa and that opens, in turn, into the gastrovascular cavity. As its name implies, the gastrovascular cavity has various functions: digestion and circulation, and as a hydrostatic skeleton. The gastrovascular cavity is divided by gastroderm-lined partitions called septa, which serve to increase the surface area of the gastrovascular cavity.

Longitudinal retractor muscles occur on the complete, or primary, septa (those that extend entirely across the gastrovascular cavity, attaching to both the gastrodermis of the outer body wall and the pharynx). The septa are paired and usually occur in multiples of twelve in the sea anemones. Incomplete septa (secondary or tertiary), when present, occur in pairs between the complete septa; they are attached at the body wall but do not connect to the pharynx. The free end of an incomplete septum is trilobed. The middle lobe contains enzyme-secreting cells and nematocysts, while the lateral lobes are ciliated and circulate fluid in the coelenteron.

The body wall consists of the usual cnidarian layers: epidermis and gastrodermis separated by mesogloea. The muscular system in the body wall consists only of gastrodermally located circular muscles, but longitudinal fibers are present in the epidermis of the tentacles, which are not served by the longitudinal septal muscles. The nervous system is the usual cnidarian nerve net.

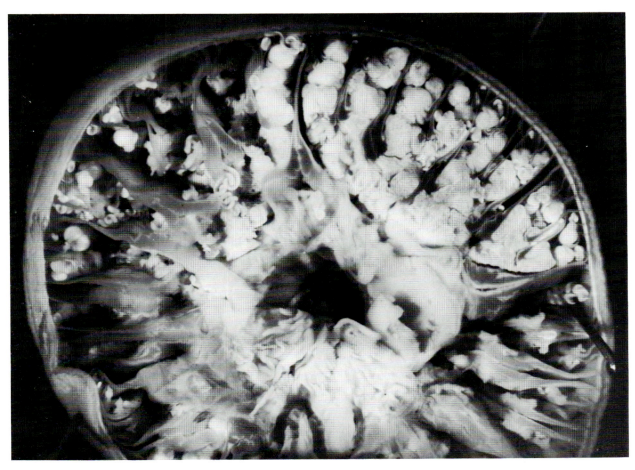

FIG. 119

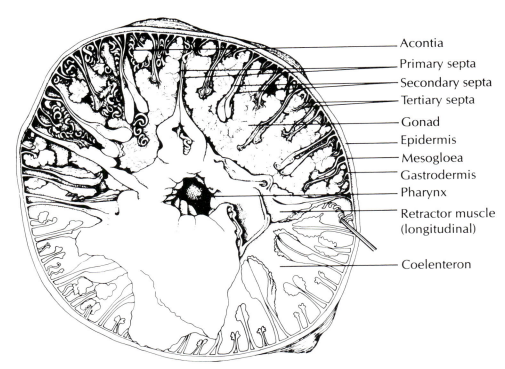

Acontia

Primary septa

Secondary septa

Tertiary septa

Gonad

Epidermis

Mesogloea

Gastrodermis

Pharynx

Retractor muscle
(longitudinal)

Coelenteron

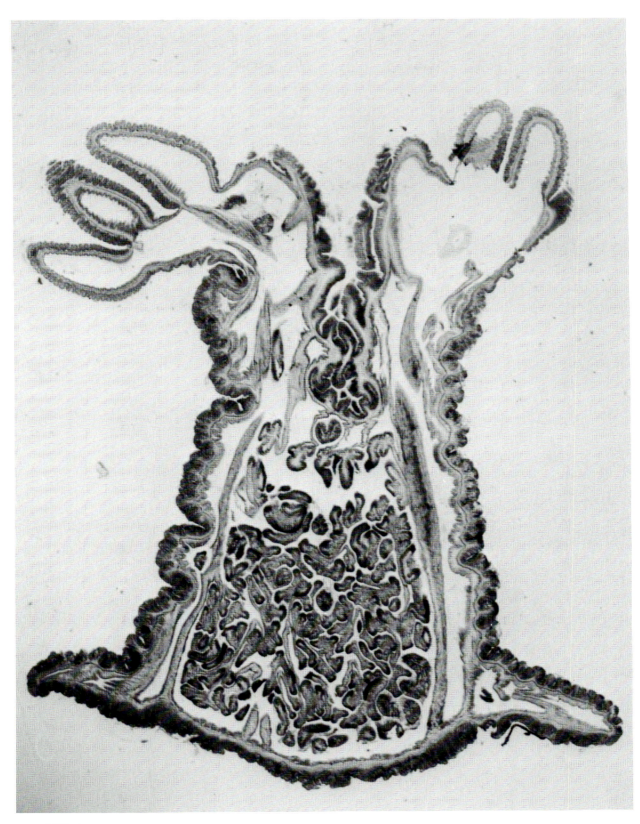

FIG. 120 *(cont.)*

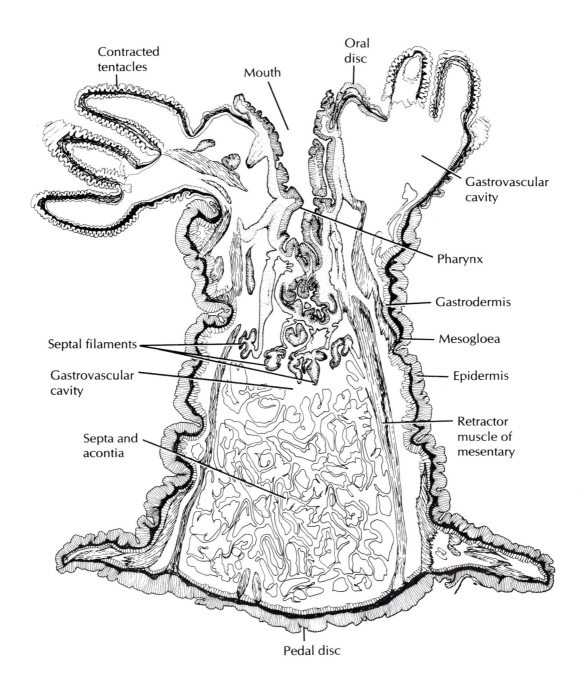

Contracted
tentacles

Mouth

Oral
disc

Gastrovascular
cavity

Pharynx

Gastrodermis

Mesogloea

Epidermis

Retractor
muscle of
mesentary

Septal filaments

Gastrovascular
cavity

Septa and
acontia

Pedal disc

FIG. 120 *(cont.)*

FIG. 121. The mouth of the actinarian sea anemones is a slit that has a ciliated tract, called a siphonoglyph, leading into the pharynx at one or both ends. The siphonoglyphs, shown here in *Metridium,* circulate water currents into the gastrovascular cavity. The siphonoglyphs impart a biradial (if two siphonoglyphs are present, as in this *Metridium*) or bilateral symmetry to an otherwise radially symmetrical animal. Some authors use this bilateral symmetry as evidence that the bilaterally symmetrical metazoa evolved from the Cnidaria, a hypothesis with which many others disagree. Some specimens of *Metridium,* for example, have only a single siphonoglyph. Thus the characteristic is variable and the symmetry it imposes is coincidental.

FIG. 122. The tentacles surrounding the oral disc of *Metridium* occur in bundles. A tentacle arrangement more typical of the actinarians—single, fairly large, evenly spaced—is shown in this specimen of *Diadumene.*

FIGS. 123 AND 124. The surface of *Metridium* is covered with a large amount of mucus. In addition, the outer surface is densely ciliated. These SEMs of a *Metridium* tentacle (Fig. 123, ×400) and outer body wall (Fig. 124 ×1000) have been treated to remove the mucus and reveal the cilia. A discharged nematocyst is present in Fig. 124.

88

FIG. 121

Siphonoglyphs

FIG. 122

FIG. 123

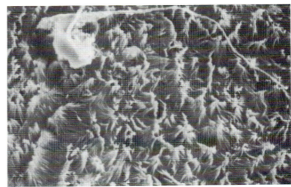

FIG. 124

FIG. 125. *Renilla,* the sea pansy, is a colonial anthozoan belonging to the subclass Alcyonaria, the polyps of which are characterized by eight pinnate tentacles. The primary polyp of *Renilla* consists of a flattened discoidal surface, the rachis, and a stalk-like peduncle that anchors the organism. Two types of secondary polyps arise from the rachis, autozooids and siphonozooids. Both are embedded in the matrix of the rachis, the coenenchyme. The autozooids are concentrated around the margins of the rachis and protrude from it. They have tentacles and are responsible for feeding. The portion of the autozooid that protrudes is called the anthocodium; the rest of the polyp is submerged in the coenenchyme. The siphonozooid polyps do not protrude from the rachis, nor do they have tentacles. The siphonozooids circulate water through the colony with well-developed siphonoglyphs. The siphonozooids depend entirely upon the autozooids for their food. Skeletal structures, the spicules, on the rachis give *Renilla* its characteristic purple color. They are secreted by amoebocytes in the mesogloea.

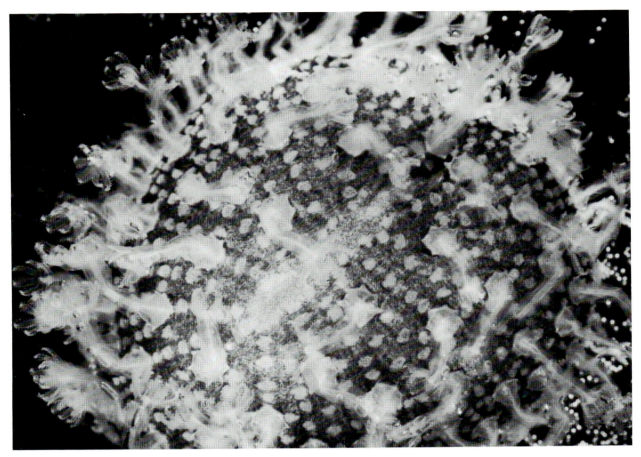

FIG. 125

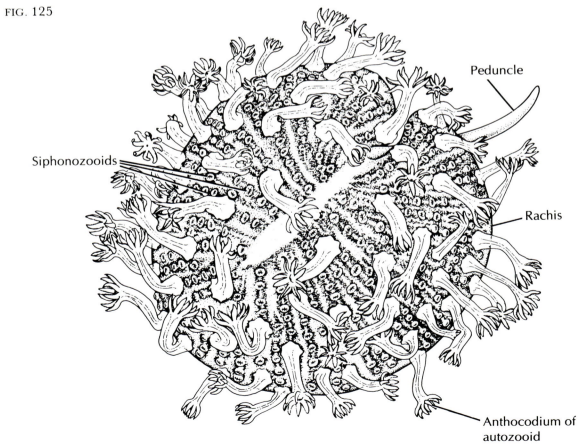

Peduncle

Siphonozooids

Rachis

Anthocodium of
autozooid

FIG. 126. The pinnate branching of the tentacles of the sea whip *Leptogorgia* are clearly evident in this anthocodium. (The polyps of *Leptogorgia* are not dimorphic.) The fine branches on the tentacles are called pinnules. A central rod, the axial rod, runs through the branches of the colony. It is composed of a proteinaceous substance called gorgonin (after the order Gorgonacea). The coenenchyme surrounding the axial rod contains calcareous spicules that give the colony its color.

FIGS. 127–129. The stony corals, although colonial, are more closely related to the sea anemone *Metridium* than to the alcyonarian *Leptogorgia. Astrangia,* shown in these three figures, is a typical example, although it is not a reefbuilder. Each polyp protrudes from a calcareous cup (theca) into which it can contract for protection (Fig. 129). The scallops of the theca, called sclerosepta, interdigitate with the base of the polyp. The skeleton is secreted by the body wall of the polyp and is composed of calcium carbonate. The tentacles in *Astrangia* are of two sizes; the large and small tentacles usually alternate (Fig. 127). The mouth is slitlike when closed and lacks siphonoglyphs (Fig. 128), but it can distend widely when *Astrangia* feeds. All of the polyps in a coral colony share a common gastrovascular cavity and nerve net. The internal anatomy is similar to that of the sea anemones.

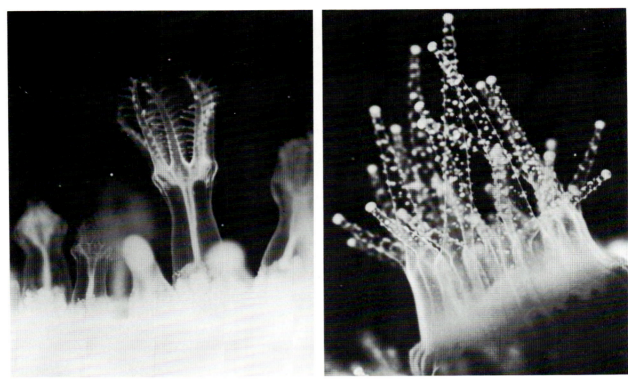

FIG. 126

FIG. 127

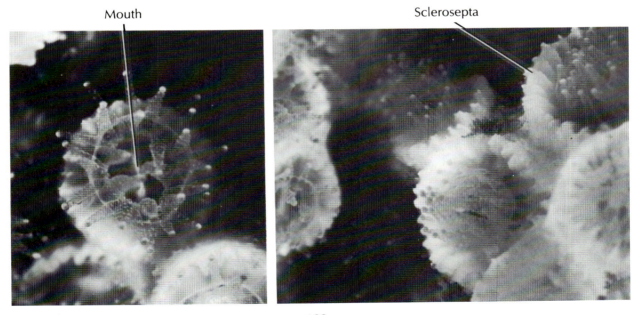

Mouth

Sclerosepta

FIG. 128

FIG. 129

FIG. 130. There are several types of nematocysts in the Cnidaria. The undischarged holotrichous isorhiza, shown sectioned here in transmission electron micrographs (TEMs) (Fig. 130a, ×8000 and b, ×8000), is a barbed nematocyst. Before it is discharged, the nematocyst is inside out within the capsule of the cnidoblast. The barbs of the thread are turned inward, much like the closed blades of a penknife. Upon appropriate stimulation, the operculum opens and the thread is ejected, turning right side out as it goes. A holotrichous isorhiza that has been discharged is shown in the SEM (Fig. 130c, ×620). The spines are now on the outside of the threads. These nematocysts are from the tentacle of the sea anemone *Corynactis californica*. (From R. N. Mariscal, "Nematocysts," in *Coelenterate Biology: Reviews and New Perspectives,* ed. L. Muscatine and H. Lenhoff [New York: Academic Press, 1974], p. 131.)

FIG. 131. Other nematocysts are adhesive or wrap around prey. These types lack spines, such as the spirocyst shown undischarged in the TEM (Fig. 131a, ×16,500) and partially discharged in the SEM (Fig. 131b, ×4470). These nematocysts are from the sea anemone *Calliactis tricolor.* (From R. N. Mariscal, ibid., p. 139.)

FIG. 132. SEM (×2640) of a nematocyst (stenotele type) torn from a tentacle of *Hydra*.

FIG. 133. LM of nematocyst batteries on an oral arm of *Cyanea*. The undischarged nematocyst threads are visible inside the capsule. A portion of a planula larva fills the upper left corner of the photograph.

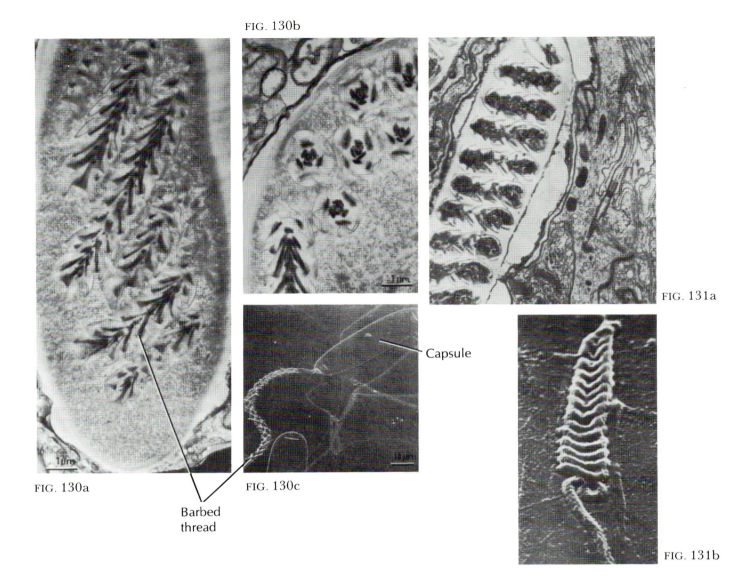

FIG. 130b

FIG. 131a

Capsule

FIG. 130a

FIG. 130c

Barbed
thread

FIG. 131b

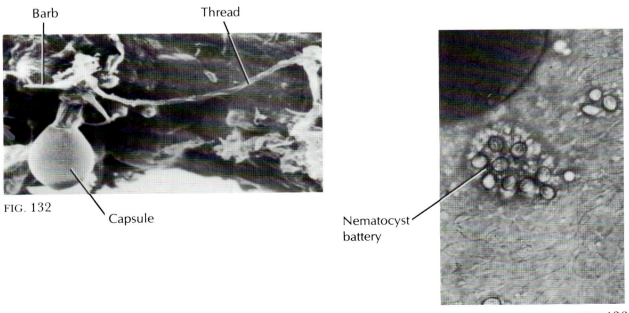

Barb

Thread

FIG. 132

Capsule

Nematocyst
battery

FIG. 133

FIG. 134. Nerve net of *Cyanea cappilata,* shown in an LM of a living preparation specially treated to remove the overlying tissue and surrounding mesogloea but leave the nerve net intact. Bipolar and multipolar neurons are evident. (*Preparation and photograph courtesy of P. Anderson, C. V. Whitney Laboratory, St. Augustine, Florida.*)

FIG. 134

SECTION IV
Taxonomic Summary

Phylum Ctenophora
CLASS TENTACULATA
Order Cydippida
Pleurobrachia
Order Lobata
Mnemiopsis

FIG. 135 The Ctenophora is a tiny phylum that is often considered to be closely related to the Cnidaria. However close the relationship may be, there are substantive morphological differences between the two groups. The cydippid ctenophores, such as *Pleurobrachia*, shown here, are representative of the standard anatomical plan. Ctenophores have two tentacles that are often extremely long and highly contractile. The tentacles are attached inside two deep pockets, the tentacle sheaths, into which they can be completely retracted. They are covered by adhesive cells, called colloblasts (see Fig. 140), that function in food capture. Locomotion is accomplished by eight comb rows or ctenes composed of fused cilia. A single, aboral statocyst (see Fig. 137) controls the relative rates of ciliary beats. An anal pore is present next to the statocyst.

Ctenophores are active predators. At times of peak abundance they may be the predominant plankton predator in localized areas such as the Chesapeake Bay.

FIG. 136. Although the phylum comprises only about 50 species, there is much morphological diversity around the standard cydippid plan. As an example, *Mnemiopsis,* a lobate ctenophore, is shown here. Its body is flattened in the tentacular plane and elongated sagittally into muscular oral lobes (see Fig. 139). The tentacles are greatly reduced and lie adjacent to the mouth. A row of short auricular tentacles lies in the auricular groove proximal to the oral lobes. Ciliary tracts run along the edges of the auricles.

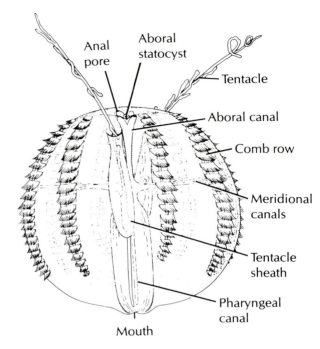

FIG. 135

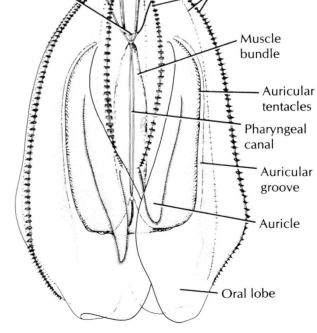

FIG. 136

FIG. 137. Close-up of the aboral statocyst of a living *Mnemiopsis.* The calcareous statoliths sit on four bundles of balancer cilia. The balancer cilia are at the ends of four ciliated furrows, which divide to form the aboral ends of the eight comb rows. The statolith and balancers are enclosed in a translucent dome. As the organism tips, the changing pressure of the statolith on the balancer cilia produces changes in the comb-row beats, which right the animal.

FIG. 138. Close-up of the comb rows of a living *Mnemiopsis.*

FIG. 139. Muscle fiber network in the oral lobe of *Mnemiopsis* traversed by a loop of the subsagittal canal of the digestive system.

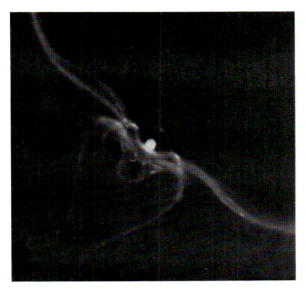

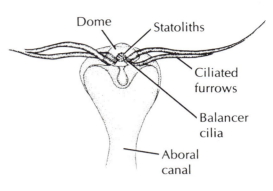

Dome Statoliths

Ciliated furrows

Balancer cilia

Aboral canal

FIG. 137

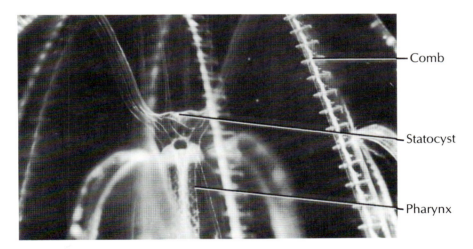

Comb

Statocyst

Pharynx

FIG. 138

Subsaggital canal   Muscle fiber network

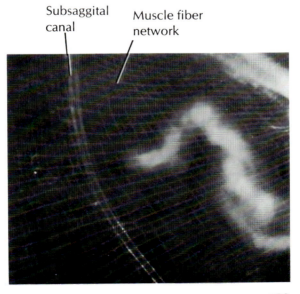

FIG. 139

FIGS. 140–142. Colloblasts are adhesive cells located on the tentacles of ctenophores. These SEMs (Fig. 140, ×660; Fig. 141, ×6270; Fig. 142, ×7000) show the colloblasts on a tentacle of *Pleurobrachia*. Unlike nematocysts, each colloblast is an entire cell. They are not fired from the tentacle, although they may be elevated in the presence of prey. The portion of the colloblast that protrudes from the tentacle is called the colloblast head. It is usually covered with many structures called refractive vesicles. These vesicles overlie another layer of vesicles called eosinophilic granules, which are the adhesive sites of the cell and which are anchored to a central spheroidal body by a fibril called a radius (see Fig. 143). The colloblast is held in place on the tentacle by a helical thread imbedded in the mesogloea by a root. The helical thread is obvious on colloblasts that have been partially freed from the tentacle (Fig. 142, ×7000).

The use of the colloblasts is under the control of neurons that synapse at the colloblast root (see Fig. 143) and may inhibit the colloblasts of a well-fed ctenophore from attaching to prey. Prey is captured when the eosinophilic granules rupture, releasing an adhesive that glues the colloblast to the prey. This process destroys the colloblast. New colloblasts are continually being formed at a site away from the tentacle and are transported by an unknown mechanism to the site of use.

FIG. 143. Schematic drawing of the fine structure in a section through a colloblast. (From J. Franc, "Organization and function of ctenophore colloblasts: An ultrastructional study," *Biol. Bull.* 155 [1978]: 537.)

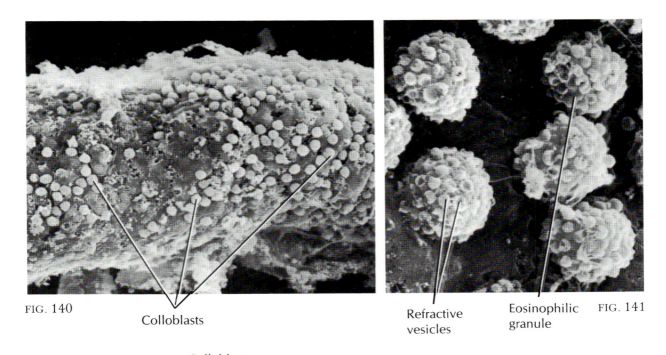

FIG. 140

Colloblasts

Refractive
vesicles

Eosinophilic
granule

FIG. 141

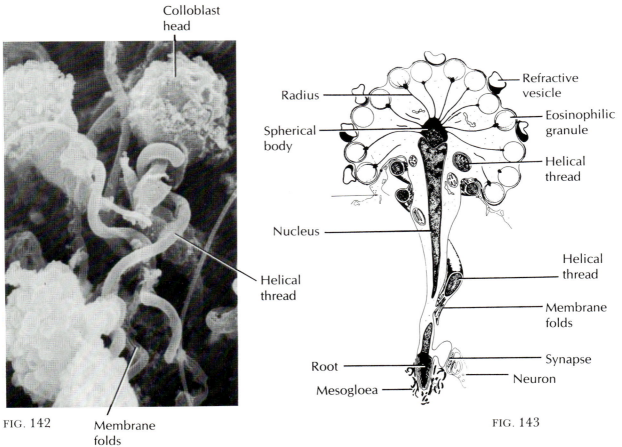

Colloblast
head

FIG. 142

Helical
thread

Membrane
folds

Radius

Spherical
body

Nucleus

Root

Mesogloea

Refractive
vesicle

Eosinophilic
granule

Helical
thread

Helical
thread

Membrane
folds

Synapse

Neuron

FIG. 143

SECTION V
Taxonomic Summary

Phylum Platyhelminthes
  CLASS TURBELLARIA
    Order Acoela
      Polychoerus
    Order Polycladida
      Leptoplana
      Stylochus
    Order Seriata
     Suborder Tricladida
      Dugesia
      Bdelloura candida
      Procotyla

FIG. 144. The flatworms (phylum Platyhelminthes) are an important phylogenetic group for several reasons. They are the most primitive of the bilaterally symmetrical metazoa. They have true organ systems. The first indications of metamerism are present and cephalization has occurred. The mesoderm, a third germ layer, is present and is expressed as a cellular parenchyma between the epidermis and gastrodermis. Two flatworm classes (not considered here), Cestoda and Trematoda, are highly specialized parasites. The acoel turbellarians, such as *Polychoerus* (shown here in a whole-mount), represent the most primitive living metazoa to proponents of the syncytial theory of metazoan origin.

According to the syncytial theory, multicellularity arose from the compartmentalization of nuclei in multinucleated ciliate protozoa. The result of this cellularization process would have been a bilaterally symmetrical ancestor similar in size, shape, and structure to an acoel flatworm. The syncytial theory has been criticized on both embryological and logical grounds. For example, that cnidarians would have been derived from an acoel flatworm seems most unlikely. Theories of evolution notwithstanding, the acoel flatworms are the least complex of the Platyhelminthes. They are found exclusively in marine environments.

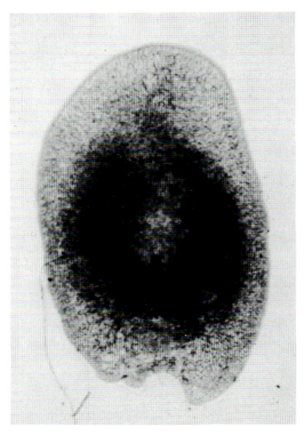

FIG. 144

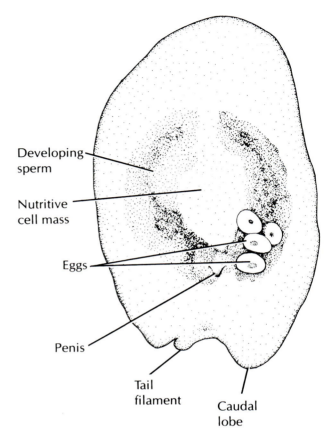

Developing sperm

Nutritive cell mass

Eggs

Penis

Tail filament

Caudal lobe

FIG. 145. The morphology of the polyclad flatworms is much more complex than that of the acoels. The digestive system is a true cavity, with a mouth opening through which a plicate pharynx is protruded during feeding. The pharynx is either ruffled, as in *Stylochus* (×30) (formerly *Planocera* and perhaps *Hoploplana*), shown here, or tubular. The intestine is a laterally branched, central cavity lying along the longitudinal axis of the worm. The nervous system, while somewhat reminiscent of the acoel plan, is more highly developed. Multiple eyespots are usually present. The polyclads have hermaphroditic reproductive systems as well as an excretory system, the latter feature lacking entirely in the acoels. Polyclads are very common inhabitants of marine benthic environments.

FIG. 146. Whole-mount (×15) of the polyclad *Leptoplana*.

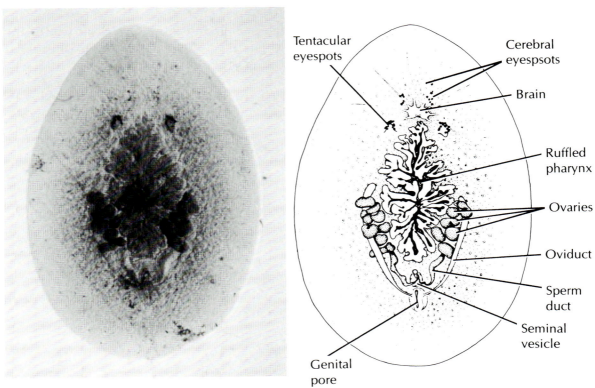

Tentacular
eyespots

Cerebral
eyespsots

Brain

Ruffled
pharynx

Ovaries

Oviduct

Sperm
duct

Seminal
vesicle

Genital
pore

FIG. 145

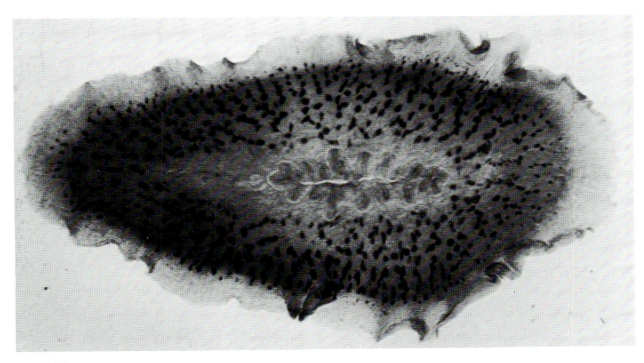

FIG. 146

FIG. 147. Many consider the triclads to be the most highly developed of the turbellarians. *Procotyla* (×17), a freshwater triclad, shown in whole-mount here, exhibits some of the typical features (see also Figs. 148 and 149). The pharynx is tubular and is everted through a central mouth. The gut has three main branches (hence, the name triclad): a single, central, anteriorly running branch and two posteriorly directed branches, one on either side of the pharynx. Each of the main branches gives rise to many laterally oriented, smaller branches, called diverticula. The triclads, like other Turbellaria, are hermaphroditic. The nervous system is much more condensed and structured than that of the acoels or polyclads. Two ventral nerve cords arise from anteriorly located ganglia (the brain). Lateral nerves branch from the ventral cords at intervals and connect with smaller longitudinal nerve cords at the periphery, giving a segmented appearance to the whole system. A well-developed protonephridial system of flame bulbs and capillaries is present, although usually difficult to see. *Procotyla*, in company with many triclads, has an adhesive disc. Such adhesive organs are used for locomotion and/or prey capture.

FIG. 148. Whole-mount (×22) of the marine triclad *Bdelloura candida*, which lives on the book gills and leg bases of the horseshoe crab *(Limulus polyphemus)*. The adhesive disc of *Bdelloura* is located posteriorly and is used to hold onto the gills of the crab. *Bdelloura* is not a parasite. It feeds on material that passes its way as the crab moves about.

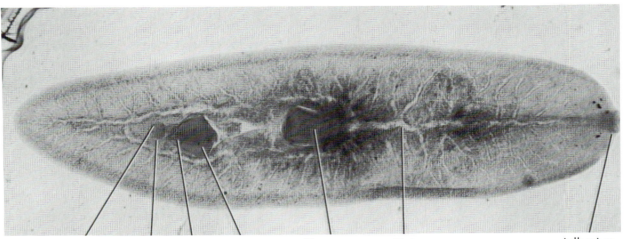

FIG. 147   Gonopore   Antrum   Penis   Penis   Pharynx   Gut   Adhesive
                                   papilla   bulb   disc

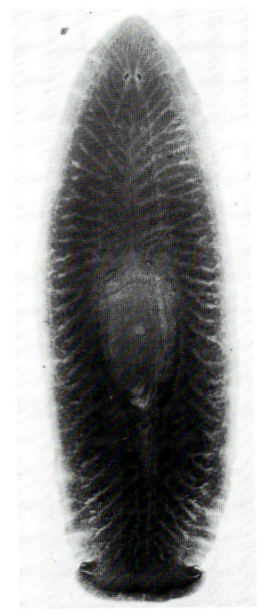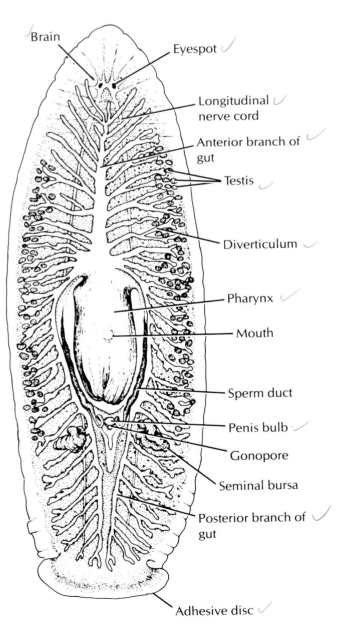

Brain

Eyespot

Longitudinal
nerve cord

Anterior branch of
gut

Testis

Diverticulum

Pharynx

Mouth

Sperm duct

Penis bulb

Gonopore

Seminal bursa

Posterior branch of
gut

Adhesive disc

FIG. 148

FIG. 149. The freshwater triclad *Dugesia* is frequently studied in the classroom as a representative platyhelminth. It is easily recognized by its triangular anterior end. *Dugesia* is unusual among the triclads for at least two reasons. First, along with sexual reproduction, *Dugesia* routinely reproduces by transverse fission, usually dividing immediately behind the pharynx. Second, *Dugesia* has excellent powers of regeneration and has been the subject of many studies in this area. *Dugesia* is also famous in a minor way for being the subject of a study in which it was claimed that memory inheritance had been demonstrated.

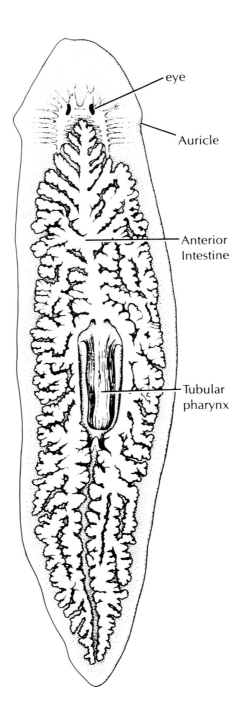

eye

Auricle

Anterior
Intestine

Tubular
pharynx

FIG. 149

FIG. 150. LM (×25) of a cross section of *Dugesia* taken at the level of the pharynx. The body wall of flatworms is fairly complex. The epidermis is ciliated (see Fig. 151), and the cilia are used in conjunction with a mucous trail for locomotion. Various secretory or gland cells are interspersed throughout the epidermis. In many flatworms, adhesive cells are concentrated in adhesive zones, for example, the adhesive discs of *Bdelloura* (Fig. 148) and *Procotyla* (Fig. 147). In *Dugesia,* an adhesive gland runs along the lateral edge of the ventral surface (see Fig. 153). Rod-shaped rhabdites, structures secreted by the gland cells and of unknown function, are also present in the epidermis. The body wall musculature lies under the epidermis and is composed of outer circular fibers, inner longitudinal fibers, and a middle section of diagonal fibers. Dorsoventral muscle fibers also run through the parenchyma. The irregularly shaped cells of the parenchyma are loosely distributed under the muscle layer and fill the space surrounding the organs.

FIG. 151. SEM (×5100) of ciliated ventral surface of *Dugesia.*

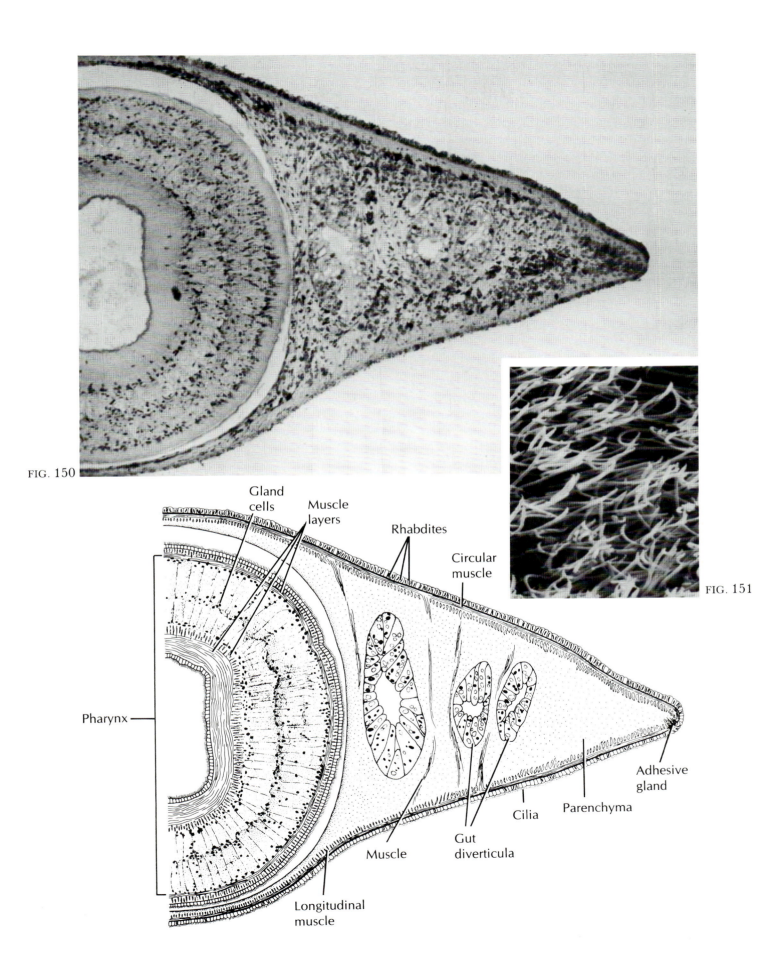

FIG. 150

FIG. 151

Gland
cells

Muscle
layers

Rhabdites

Circular
muscle

Pharynx

Muscle

Gut
diverticula

Cilia

Parenchyma

Adhesive
gland

Longitudinal
muscle

FIG. 152. Higher-magnification LM (×440) (cross section) of the parenchyma of *Dugesia*. The excretory system, which is very difficult to demonstrate histologically, consists of a series of branched, blind-ended tubules. At the termination of each branch (capillary) is a flagellated cell called a flame cell. Each branched tubule is called a protonephridium. A segment of a capillary and what might be a flame cell are visible here.

FIG. 153. LM (×350) of the adhesive gland of *Dugesia* (cross section).

FIG. 154. Higher-magnification LM (×320) (cross section) of the ventral surface of *Dugesia*. One of the ventral nerve cords and some of the circular musculature are evident. The walls of the digestive diverticula consist of gland cells and phagocytic cells.

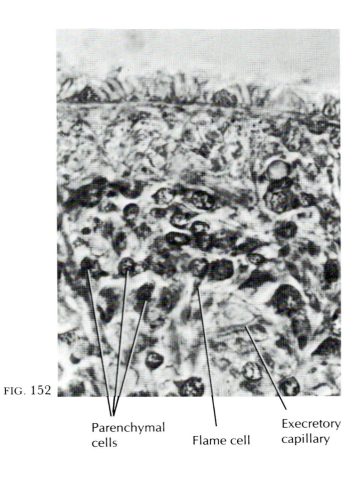

FIG. 152

Parenchymal
cells   Flame cell   Execretory
                                capillary

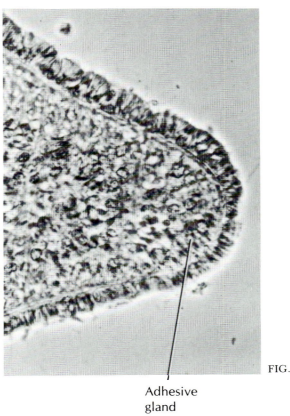

FIG. 153

Adhesive
gland

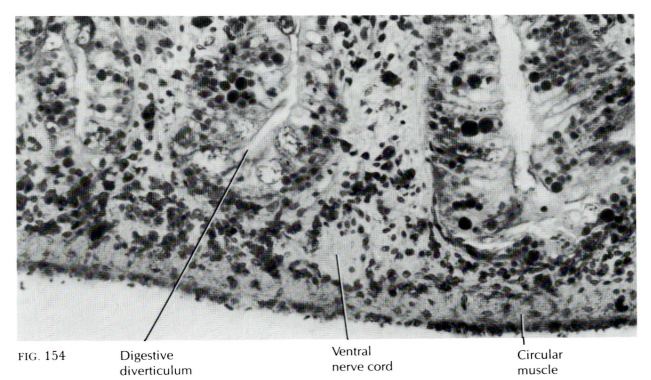

FIG. 154   Digestive   Ventral   Circular
                diverticulum   nerve cord   muscle

SECTION VI
Taxonomic Summary

SOME MINOR PHYLA
Phylum Nemertinea (Rhynchocoela)
   CLASS ANOPLA
     Order Heteronemertea
       Cerebratulus lacteus

ASCHELMINTHES (PSEUDOCOELOMATES)
Phylum Nematoda
       Turbatrix aceti
Phylum Rotifera
   CLASS DIGONATA
     Order Bdelloidea
       Philodina
   CLASS MONOGONATA
     Order Ploima
       Monostyla

LOPHOPHORATE COELOMATES
Phylum Brachiopoda
   CLASS INARTICULATA
     Lingula

MINOR PROTOSTOME GROUPS
Phylum Sipuncula
       Phascolosoma (Golfingia) gouldii
Phylum Echiura
       Urechis caupo

FIG. 155. Worms in the phylum Nemertinea are closely allied to the flatworms. In fact, some taxonomic schemes have included the nemerteans within the phylum Platyhelminthes. There are many morphological similarities: ciliated epidermis (even rhabdites in some species), parenchyma, protonephridia with flame bulbs, and an acoelomate body plan. However, nemerteans differ from flatworms in several other characteristics: closed circulatory system, more advanced nervous and muscular systems, complete digestive system, and a unique hydrostatic organ (the rhyncocoel).

The rhyncocoel, a fluid-filled body cavity, functions to evert the worm's feeding structure, the proboscis. Thus, as a hydrostatic skeleton, the rhyncocoel is considered phylogenetically important because it is functionally comparable to the coelom of higher animals. In spite of the nemerteans' advances over the flatworms in body organization and the appearance of a true hydrostatic organ reminiscent of a coelom, most authors agree that these worms are an evolutionary byway well off the main lines leading to the coelomates.

Cerebratulus, shown in dissection here, is itself an aberration among the nemerteans (most ribbon worms are much smaller) but it is large enough to be dissected easily and is thus often used as a representative of the group.

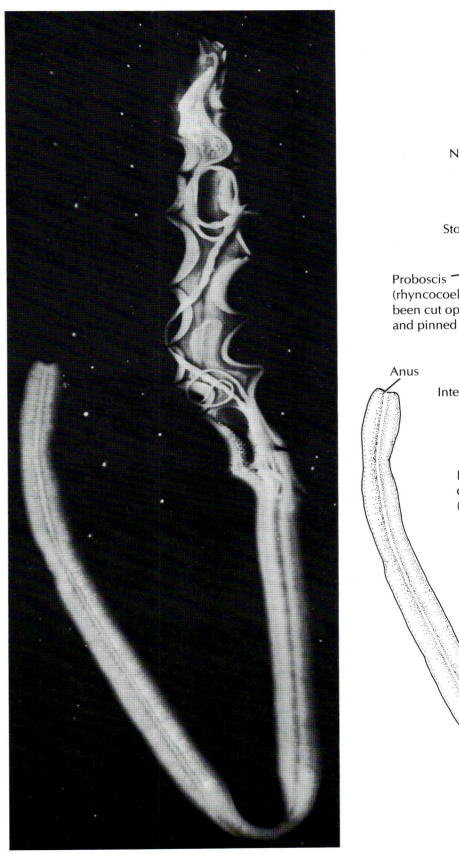

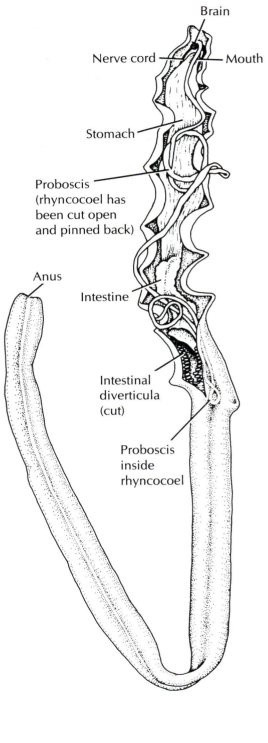

Brain

Nerve cord — Mouth

Stomach

Proboscis
(rhyncocoel has
been cut open
and pinned back)

Anus

Intestine

Intestinal
diverticula
(cut)

Proboscis
inside
rhyncocoel

FIG. 155

FIG. 156. Dissection of a living *Cerebratulus* that had been narcotized with menthol crystals. The anterior end is shovel-shaped, as befits a burrowing worm. The proboscis is everted through a pore at the extreme anterior end. The rhyncocoel leading to the proboscis pore is visible through the semitransparent body wall. Intestinal diverticula are also visible through the more posterior portions of the body wall, shown here in a loop around the anterior end.

FIG. 157. Close-up of the dissected anterior end of *Cerebratulus*. The rhyncocoel has been cut and pinned open. One lobe of the brain is visible. The outline of the mouth, which opens on the underside, can also be seen.

FIG. 158. Close-up of a more posterior portion of *Cerebratulus*. Part of the rhyncocoel is intact and surrounds the proboscis. The dissection extends from the rhyncocoel down into the intestinal cavity and cuts across several of the tubular diverticula.

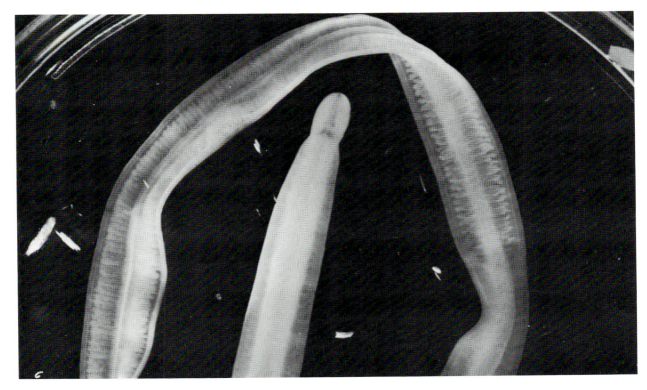

FIG. 156

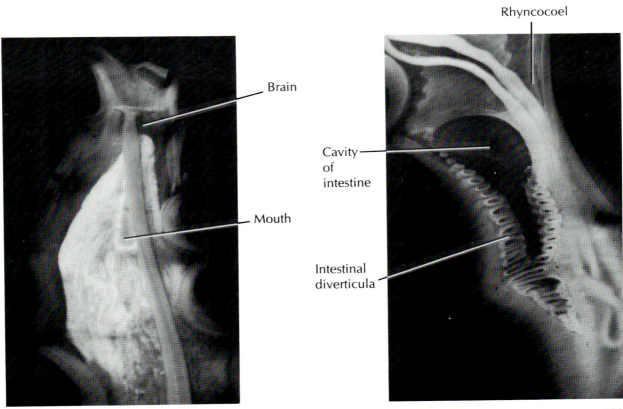

FIG. 157

Brain

Mouth

Rhyncocoel

Cavity
of
intestine

Intestinal
diverticula

FIG. 158

FIG. 159. The Nematoda is one of several phyla often placed together under the group name Aschelminthes. The morphologies of the various aschelminths are quite different, but they all have in common the characteristic pseudocoelom. This fluid-filled body cavity arises from the embryonic blastocoel and differs from a true coelom in its lack of a peritoneal covering. The organs of pseudocoelomates lie free in the pseudocoel; there are no mesenteries to support them. In the nematodes—for instance, the vinegar eel *Turbatrix aceti,* shown in this LM (×160)—the pseudocoel functions as a hydrostatic skeleton.

The Nematoda is a large phylum with both free-living and parasitic species. Nematodes are found in virtually all habitats and, where they occur, they are often extremely abundant: there have been estimated to be 527,000 nematodes per acre in the top 3 inches of beach sand, and 36 species and 1,074 individuals in 6.7 cubic centimeters of mud. Nematodes often occupy habitats unexploited by other species. *Turbatrix* lives in vinegar. Another species lives exclusively in the straw mats that are placed beneath beer kegs in Europe. Still others are extremely tolerant of environmental extremes. For example, some can survive in liquid nitrogen, others in hot springs.

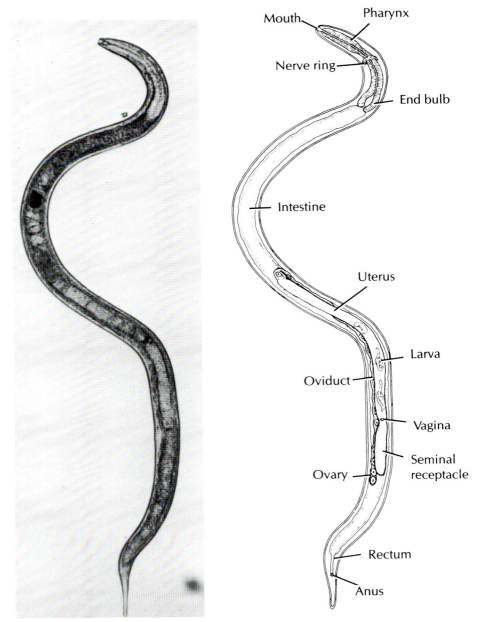

Mouth

Pharynx

Nerve ring

End bulb

Intestine

Uterus

Larva

Oviduct

Vagina

Ovary

Seminal
receptacle

Rectum

Anus

FIG. 159

FIGS. 160 AND 161. Nematodes can tolerate extreme environmental conditions because they are protected by an external cuticle. The cuticle consists of several layers, most of which contain collagenous fibers. The cuticle may be smooth or sculptured as can be seen in these LMs (Fig. 160, ×170; Fig. 161, ×320) of an unidentified marine nematode. (The rings are in the cuticle.) In this species, the cuticle forms anterior bristles that are probably sensory.

FIGS. 162 AND 163. Higher magnification LMs (Fig. 162, ×320; Fig. 163, ×520), using interference optics, of the anterior and posterior ends of *Turbatrix aceti*.

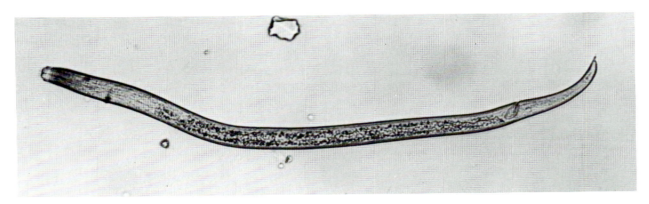

FIG. 160

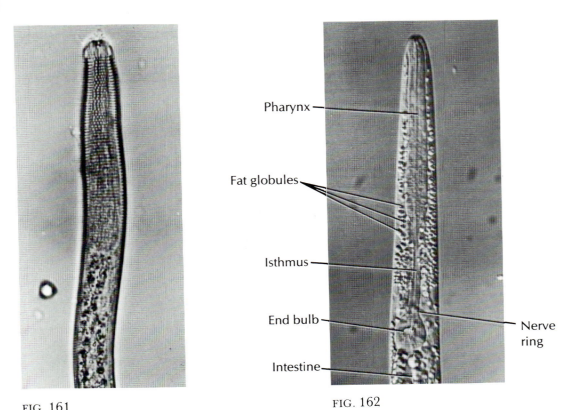

FIG. 161

Pharynx

Fat globules

Isthmus

End bulb

Intestine

Nerve ring

FIG. 162

Ovary

Rectum

FIG. 163

FIG. 164. Phylum Rotifera is also in the aschelminth group. Rotifers are tiny animals characterized by a ciliated anterior corona or wheel organ. The body is made up of sixteen pseudosegments—"pseudo" because the metamerism is of the cuticle only and does not extend internally to the organs. *Philodina,* shown in this LM (×480), is a typical rotifer.

The rotifers exemplify a classical aschelminth characteristic: eutely. That is, there is a constant number of cells or nuclei in the organs of every animal in each species. Several tissues appear to be truly syncytial. The epidermis, stomach, and germovitellarium have all been reported to contain a constant number of nuclei but no recognizable cells.

Most rotifers are female. *Philodina* and other bdelloid species have no males at all. The females produce eggs that develop parthenogenetically (i.e., without being fertilized). In those species that have both sexes, the males are usually much smaller than the females, are short-lived, and, like the females, are haploid.

Rotifers can be extremely abundant in fresh water, and they occasionally appear in temporary puddles. Several marine species and a few terrestrial species (usually associated with mosses) have been described. Some rotifers show physiological tolerances comparable only to those of the nematodes. Many species can tolerate desiccation, and freezing or hot-spring temperatures.

FIG. 165. LM (×1390) of the pharynx, or mastax, of *Philodina.* This structure, a muscular organ lined with cuticularized plates called trophi that function in mastication, is characteristic of the rotifers.

FIG. 166. Higher magnification LM (×780) of the spurs of *Philodina.* Many rotifers have toes that they use to attach to the substrate. Some rotifers attach preferentially to particular substrates, in some instances to a particular species of crustacean or aquatic plant. The toes in *Philodina* are reduced; the spurs are the more obvious feature.

FIG. 167. The morphology of *Monostyla,* a rotifer in the order Ploima, is markedly different from that of *Philodina.* This specimen, photographed with interference optics (×300), is contracted inside its lorica. *Monostyla* has a single toe.

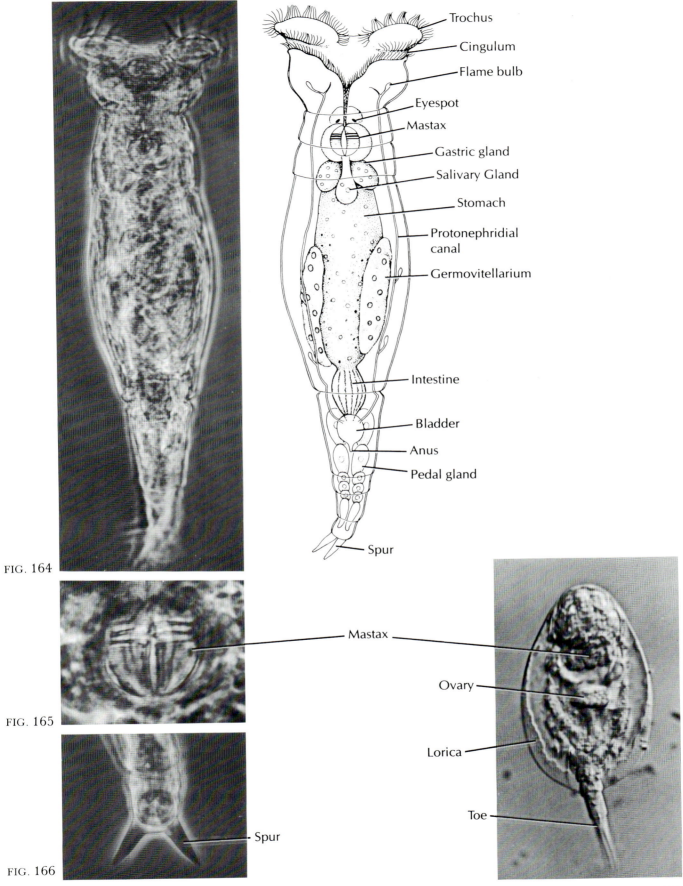

Trochus

Cingulum

Flame bulb

Eyespot

Mastax

Gastric gland

Salivary Gland

Stomach

Protonephridial canal

Germovitellarium

Intestine

Bladder

Anus

Pedal gland

Spur

FIG. 164

Mastax

Ovary

Lorica

Toe

FIG. 165

Spur

FIG. 166

FIG. 167

FIGS. 168 AND 169. Phylum Brachiopoda is a small phylum of animals that is grouped with two other phyla, Bryozoa and Phoronida, as the lophophorate coelomates. These animals have a true coelom and have in common a characteristic feeding structure, the lophophore. The lophophore is a food-gathering organ encircling the mouth and covered with tentacles. Cilia on the tentacles create a current that draws water across the lophophore. Suspended food material is trapped by mucus and transported to the mouth along ciliary tracts on the tentacles in a manner analogous to the ciliary feeding mechanism of the gills in bivalve molluscs.

The lophophorates are peculiar in that they display a mixture of protostome and deuterostome embryological traits. The brachiopods formerly were classified with the molluscs because of their bivalved shells, but, when the shell is opened (Fig. 169), the anatomy is clearly non-molluscan.

Lingula (Fig. 168) is a stalked brachiopod; many species lack stalks.

Brachiopods were particularly abundant during the Paleozoic and Mesozoic eras, but most species are now extinct. *Lingula* may be one of the oldest genera of animals with species still extant.

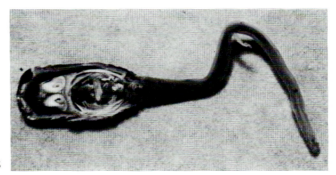

FIG. 168

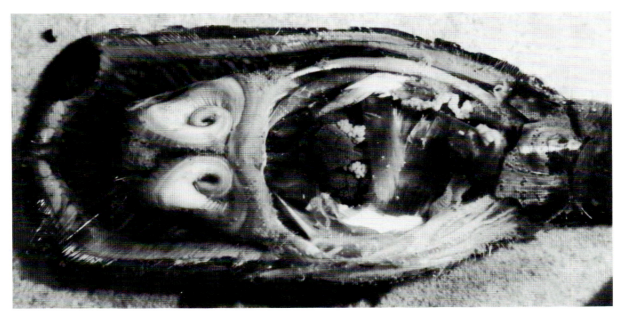

FIG. 169

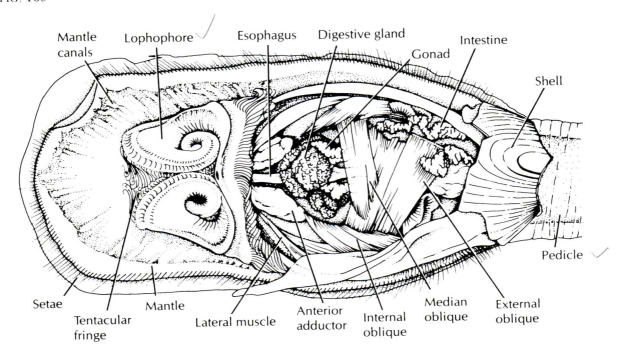

Mantle canals   Lophophore ✓   Esophagus   Digestive gland   Intestine

Gonad   Shell

Setae   Mantle   Anterior adductor   Internal oblique   Median oblique   External oblique

Tentacular fringe   Lateral muscle   Pedicle ✓

FIG. 170. The Sipuncula is a small phylum of coelomate worms that seem to be fairly close relatives of the annelids, although they are not segmented. Sipunculid anatomy, demonstrated in this dissection of *Phascolosoma gouldii,* is very simple. Sipunculids are active burrowers and deposit feeders; several species are rock borers. Although the phylum contains relatively few species, where they occur, sipunculids may be extremely abundant.

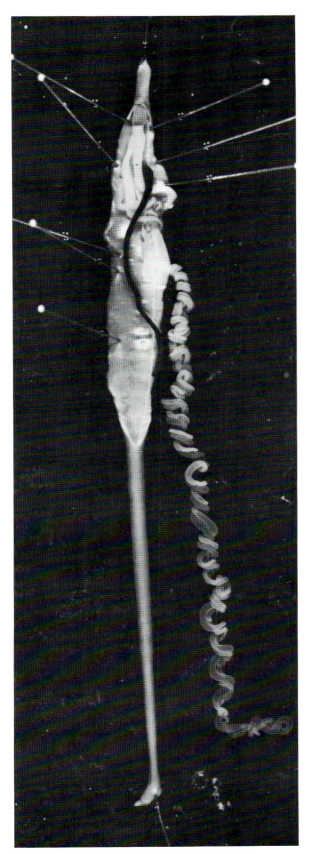

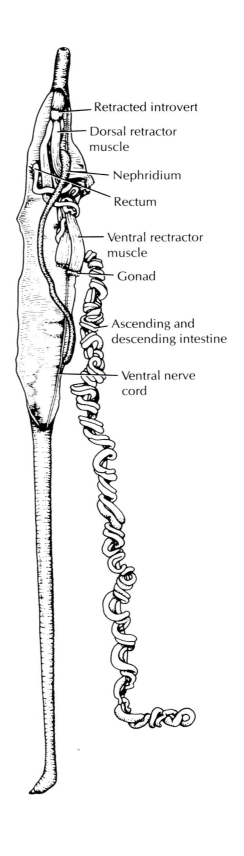

Retracted introvert

Dorsal retractor
muscle

Nephridium

Rectum

Ventral rectractor
muscle

Gonad

Ascending and
descending intestine

Ventral nerve
cord

FIG. 170

FIG. 171. The Echiurida is another small phylum of coelomate worms. They are similar to the sipunculids in size and shape, and originally both were placed in the same phylum. A transient metamerism occurs in the embryology of some echiuran species, indicating a phylogenetic similarity to the annelids. This dissection of *Urechis caupo* demonstrates its relatively simple internal anatomy.

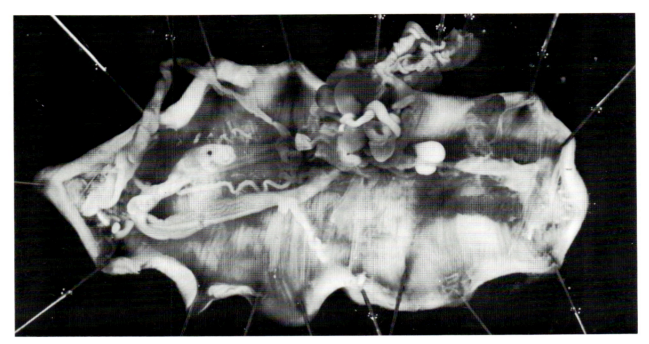

FIG. 171

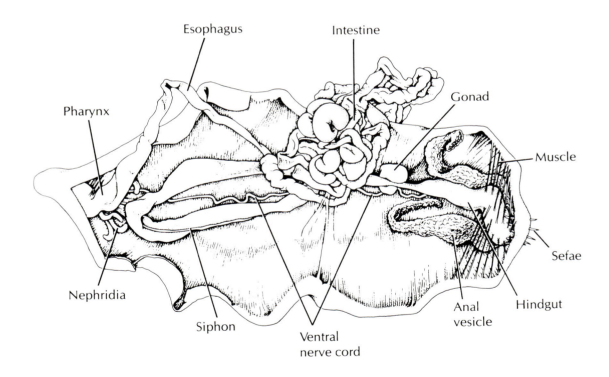

FIGS. 172–175. Over the years, the archiannelids have been grouped variously as a phylum, a class, or a polychaete subclass or order. The taxonomic difficulties stem from the unique morphology of the archiannelid species. "Archi-" implies that the organisms are primitive, but many of their external features seem to be adaptations to the interstitial environment in which they live.

Nerilla antennata, shown here in SEM (Fig. 172, ×150) and LM (Fig. 173, ×200), differs from many of the archiannelid species because it exhibits some polychaetelike anatomy. For example, *Nerilla* is obviously segmented externally, has parapodia bearing both setae and cirri (Figs. 172, and 175 ×600), and has palps and tentacles on the prostomium. Many of the other archiannelid genera lack these features. However, in common with many archiannelids, *Nerilla* has external ciliated bands that surround the mouth and extend longitudinally along the ventral surface (Fig. 174, ×540). Less-developed patches of cilia occur on the posterior dorsal surface of each segment (Fig. 172). Such external ciliation is generally considered to be a primitive characteristic among the annelids.

FIG. 172

Setae   Cirrus   Palp

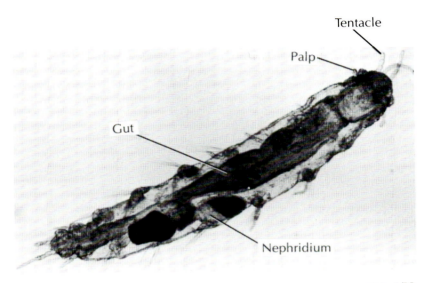

Tentacle

Palp

Gut

Nephridium

FIG. 173

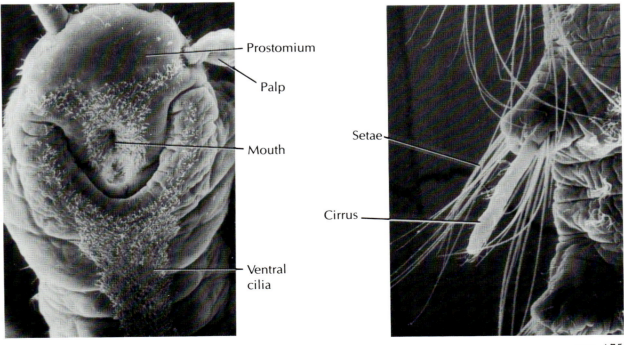

Prostomium

Palp

Mouth

Ventral
cilia

Setae

Cirrus

FIG. 174

FIG. 175

FIGS. 176–178. The body of polychaete worms of subclass Errantia is characteristically composed of similar somites; it is not divided into distinctive regions. Although the name of the subclass implies that these worms are free living, many species live in tubes or burrows, and their external morphology often reflects this. *Nereis virens* (Fig. 176) is a benthic-crawling species that occasionally swims. On its head (consisting of prostomial and peristomial segments) are several well-developed tentacles, as well as palps, cirri, and eyespots. The parapodia are large, well-developed structures (see Fig. 181) with prominent noto- and neuropodial lobes, cirri, and setae.

It might be expected that many of these delicate, fleshy external appendages would be damaged in a species that spends most of its time burrowing; indeed, they are less well developed in the burrowing forms. In *Glycera dibranchiata* (Fig. 177), for example, a species that burrows in sand flats, the anterior end is pointed and dorsoventrally flattened. The fleshy portions of the parapodia are reduced, but the parapodial structures associated with burrowing, the setae, are well developed.

Many of the errant polychaetes have a protrusile, rapidly extensible pharynx that is armed with chitinous jaws. The omnivorous *Nereis virens* has two such jaws (Fig. 176b) with which it slices up algae or captures invertebrates. *Glycera* is a raptorial feeder; it captures prey, which cross the openings of its burrow gallery, with a protrusile proboscis (Fig. 177b) armed with four jaws (Fig. 178). A gland attached to each jaw releases a poison into the prey. The hydrostatic skeleton (coelom and longitudinal body wall muscles) in *Glycera* can evert the long proboscis with great force.

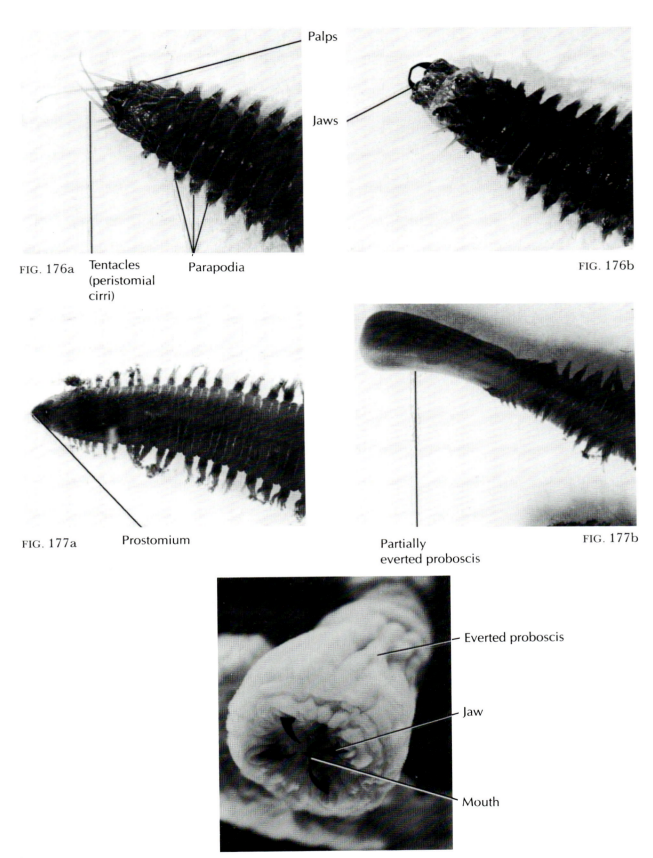

Palps

Jaws

FIG. 176a   Tentacles   Parapodia   FIG. 176b
            (peristomial
            cirri)

FIG. 177a   Prostomium   Partially   FIG. 177b
                                                 everted proboscis

                                                                  Everted proboscis

                                                                  Jaw

                                                                  Mouth

FIG. 178

FIG. 179. Another errant species, *Diopatra cuprea,* lives in a chimneylike, permanent tube that partially protrudes above the sediment. The tube is made of debris cemented with mucus. Since *Diopatra* does not leave its tube (although it partially emerges to capture prey) and since the tube is open only at one end, the sensory and respiratory structures at the anterior end of the worm are well developed. The well-developed setae on the posterior segments anchor *Diopatra* in place in the tube or else are used in rapidly levering the animal backward or forward within the tube.

FIG. 180. The scale worms are surface-dwelling, errant polychaetes covered by paired scales called elytra. Each elytron, shown here on *Lepidonotus squamatus,* is attached to the body wall by a stalk. In some species there is one pair of elytra per segment; in other species the elytra are on alternate segments. In some species, the elytra bioluminesce.

FIG. 181. The parapodia are a diagnostic feature of the polychaetes. They have various functions: swimming, crawling, burrowing, respiration, and food capture. There is one pair of parapodia per segment. The parapodium is a biramous appendage with two lobes, the dorsal notopodium and the ventral neuropodium. The lobes are supported internally by chitinous rods called acicula (see Fig. 183). The parapodium shown here is on *Nereis virens* and demonstrates the basic parapodial anatomy. However, parapodial structure varies tremendously from species to species.

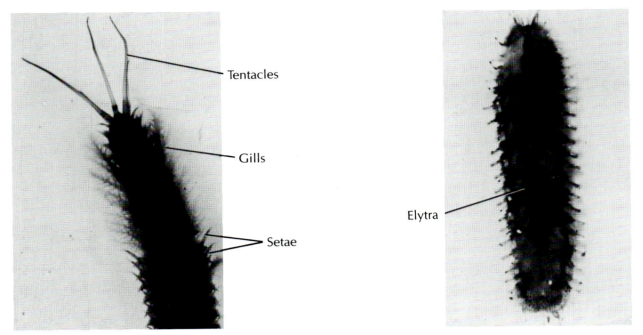

Tentacles

Gills

Setae

FIG. 179

Elytra

FIG. 180

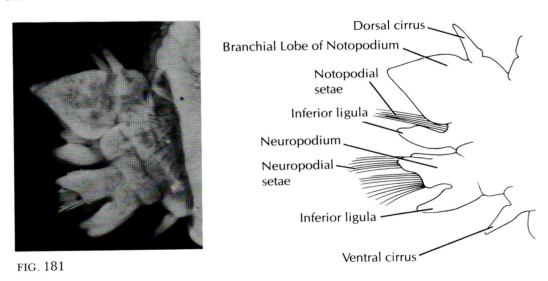

FIG. 181

Dorsal cirrus

Branchial Lobe of Notopodium

Notopodial
setae

Inferior ligula

Neuropodium

Neuropodial
setae

Inferior ligula

Ventral cirrus

FIG. 182. The annelids are true coelomates. The organ systems are repeated in series within the body; there is usually one pair of structures per segment. This serial repetition is especially obvious in the nervous, circulatory, excretory, and muscular systems. *Nereis virens,* shown here in dorsal dissection, is typical. The segments, or somites, are at least partially divided from adjacent somites by septa.

In some species, for example, *Lumbricus* (see Fig. 195), the septa are highly muscularized structures that completely segregate the somites into discrete hydrostatic units. The complete septa facilitate fine control over the hydrostatic skeleton of the burrowing worm. However, in many species such as *Nereis,* the septa are not complete and their use must be largely to assist the mesentaries in organ support.

FIG. 182

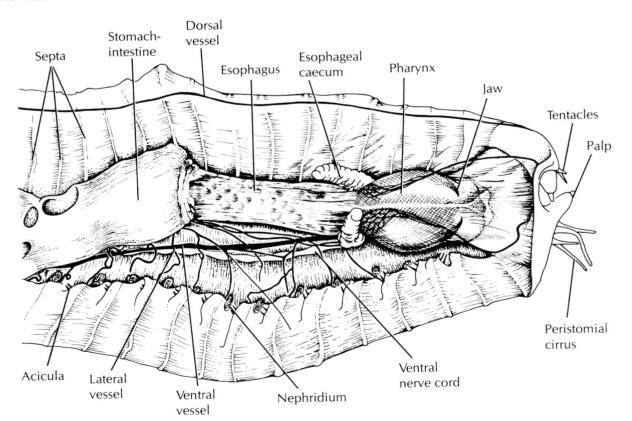

Septa

Stomach-intestine

Dorsal vessel

Esophagus

Esophageal caecum

Pharynx

Jaw

Tentacles

Palp

Peristomial cirrus

Ventral nerve cord

Nephridium

Ventral vessel

Lateral vessel

Acicula

FIG. 183. LM (×25) of cross section through a somite of *Nereis virens.*

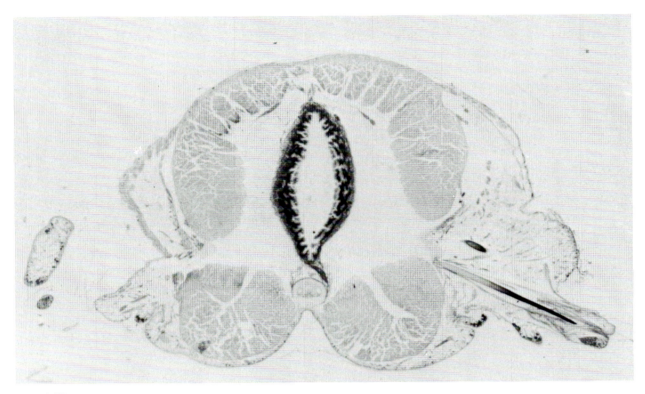

FIG. 183

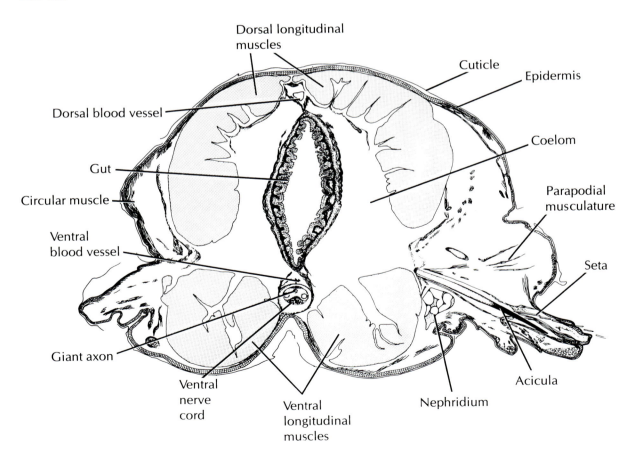

Dorsal longitudinal muscles

Cuticle

Epidermis

Dorsal blood vessel

Coelom

Gut

Circular muscle

Parapodial musculature

Ventral blood vessel

Seta

Giant axon

Acicula

Ventral nerve cord

Ventral longitudinal muscles

Nephridium

FIG. 184. Members of subclass Sedentaria are usually confined to tubes or burrows. The body may be regionally differentiated. The parapodia are reduced and lack compound setae and acicula. There are no teeth or jaws, but there are often elaborate feeding or respiratory structures associated with the head, such as the highly extensible tentacles and gills of *Amphitrite ornata* (family Terebellidae) shown here. Many of the sedentary species are filter- or deposit-feeders.

Amphitrite lives near the low tide mark in a mucus-lined burrow. Its body has three distinct regions: the head is the fused prostomium and peristomium, with attendant appendages; the thorax consists of segments with setae-bearing notopodia; and the posterior-most section lacks setae.

FIG. 185. The body of *Chaetopterus variopedatus* (family Chaetopteridae), shown here removed from its U-shaped parchment tube, is strikingly regionalized. *Chaetopterus* feeds by circulating water through its tube by moving the fan notopodia in a synchronous wave. The effective stroke of these fans can be reversed to pump water in either direction. *Chaetopterus* periodically changes direction inside the tube as well. The water current entering the tube is passed through a mucus net secreted by the food cup and suspended from the tips of the aliform notopodia of the tenth segment. Food material is trapped in the net—one of the most efficient biological filters known—which retains particles as small as 40 angstroms in diameter. As the filter fills, it is wound up and deposited on an anterior-running, middorsal ciliated groove. The cilia conduct the net to the mouth, where the food is ingested.

Eggs shed by *Chaetopterus* into seawater enter an arrested metaphase until fertilization occurs. This characteristic has been very useful to developmental and cell biologists interested in the details of cell division.

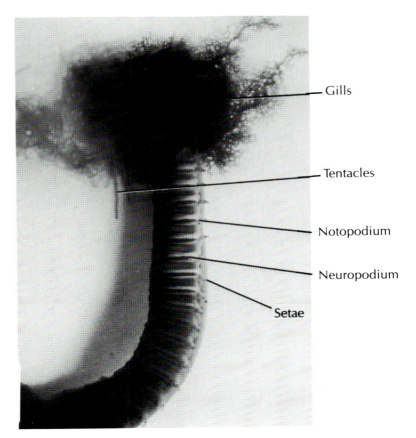

FIG. 184

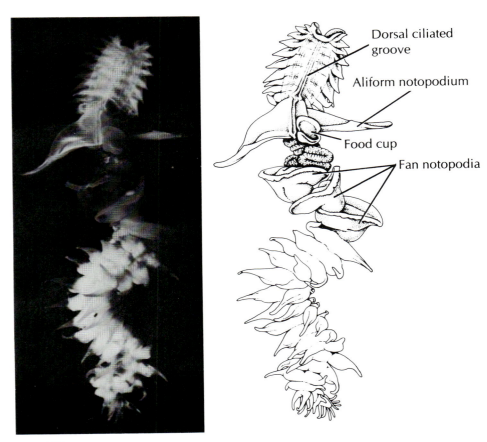

FIG. 185

FIGS. 186–189. The serpulid polychaetes are another family of tube-dwelling Sedentaria. Their tubes are calcareous encrustations such as those covering the *Mercenaria* shell in Fig. 186. The tube is fashioned by the peristomium which molds the secretions of two calcium carbonate secreting glands.

The serpulids are filter feeders. The filter is a funnel-shaped fan of pinnate branches, the radioles, that are formed from the prostomium. The radioles of a living *Hydroides dianthus* are shown partially expanded in Fig. 187. When feeding, the radioles open like a fan into the surrounding water. Cilia on each radiole (see Fig. 192) draw water through the structure and food material is filtered out, trapped in mucus, and conducted to the mouth via ciliated tracts. One of the radioles is modified into a structure called the operculum. When the fan is retracted into the tube, the operculum is drawn in last and plugs the opening of the tube.

The portion of the body of *Hydroides* that always remains in the tube has a typical sedentarian morphology (Fig. 188, ×8), including reduced parapodia and small setae. The parapodia of the segments at the mouth of the tube (Fig. 189, SEM, ×18) are more highly developed.

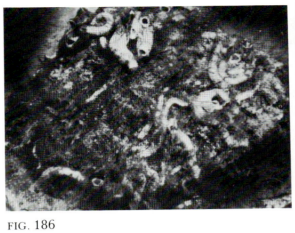

FIG. 186

FIG. 187

Operculum   Radiole

Tube

FIG. 188

Setae   Radioles   Operculum

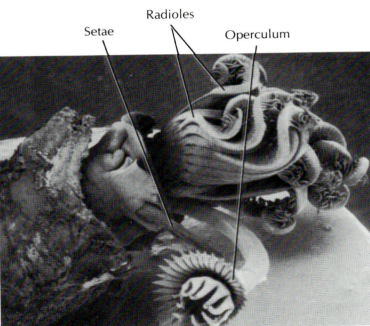

FIG. 189

FIGS. 190 AND 191. *Spirorbis spirillum* is another tube-dwelling serpulid, shown here removed from its calcareous tube (Fig. 190, SEM, ×90). The collar, which molds the tube, and its associated setae are obvious. Immediately below the notopodial setae are pairs of gills (Fig. 191, SEM, ×230).

FIG. 192. The radioles of the serpulids are highly ciliated, as can be seen in this SEM (×215) of the *Spirorbis* fan.

FIG. 193. *Pectinaria gouldii* (family Pectinariidae) is a sedentary polychaete that makes a tube—open at both ends—from similar-sized sand grains. The worm lives in sand with its head end down and carries its tube with it as it moves around in the sediments. *Pectinaria* can lengthen or repair the tube but is unable to entirely rebuild it, a task that only a juvenile worm can accomplish. *Pectinaria* is a deposit feeder. The very heavy, golden-colored setae on its head are used both for digging and as an operculum to cover the opening of the tube when the worm withdraws.

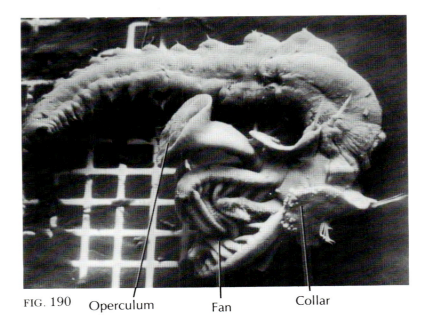

FIG. 190   Operculum   Fan   Collar

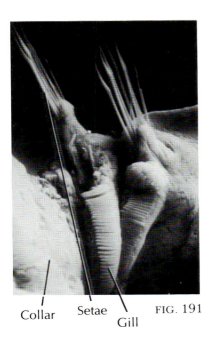

Collar   Setae   FIG. 191
Gill

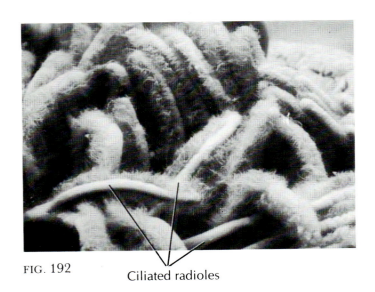

FIG. 192   Ciliated radioles

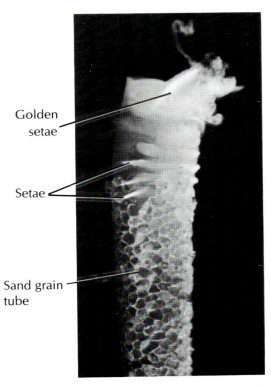

Golden
setae

Setae

Sand grain
tube

FIG. 193

FIG. 194. Dorsal dissection of the lugworm *Arenicola cristata* (family Arenicolidae), a large, sedentary polychaete that lives in a U-shaped, mucus-lined tube. Peristaltic waves passing anteriorly along the body irrigate the tube. *Arenicola* is a deposit feeder, using a very muscular pharynx to ingest sediments. The body is divided into four regions externally: the head consists of the prostomium, peristomium, and first segment; the next six segments bear reduced parapodia with setae; the next eleven segments have reduced parapodia and gills; and the tail has a variable number of segments with neither parapodia nor gills. All of the segments show external annulations.

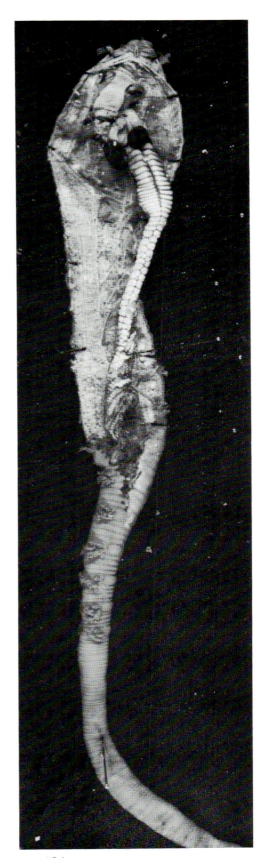

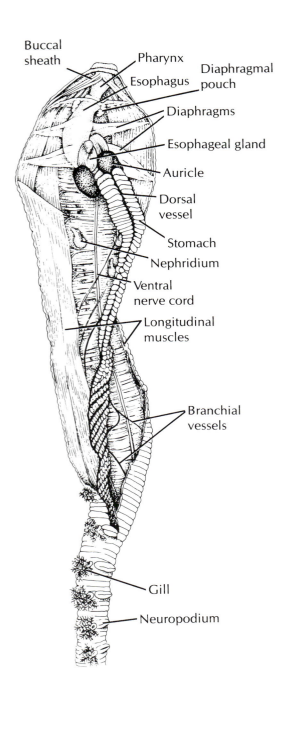

Buccal
sheath

Pharynx

Diaphragmal
pouch

Esophagus

Diaphragms

Esophageal gland

Auricle

Dorsal
vessel

Stomach

Nephridium

Ventral
nerve cord

Longitudinal
muscles

Branchial
vessels

Gill

Neuropodium

FIG. 194

FIG. 195. The external morphology of worms in class Oligochaeta is much simpler than that of the polychaetes. They have no parapodia, but setae, which aid in burrowing, are usually present. Certain segments of adult oligochaetes are noticeably thicker than adjacent segments. A glandular area of these thicker segments, called the clitellum, secretes mucus and the cocoon during reproduction (see Fig. 198). The position of the clitellum along the body varies from worm to worm.

Lumbricus terrestris, shown in ventral dissection here, is the best known of the oligochaete earthworms. Many other oligochaete species live in terrestrial, freshwater, and marine environments. In general, the terrestrial species are larger; *Megascolides,* the giant Australian earthworm, reaches lengths of more than 3 meters.

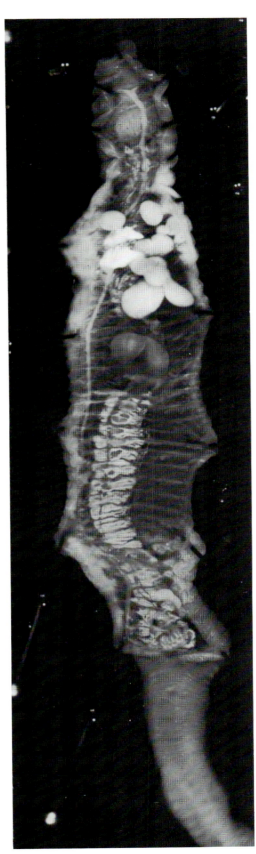

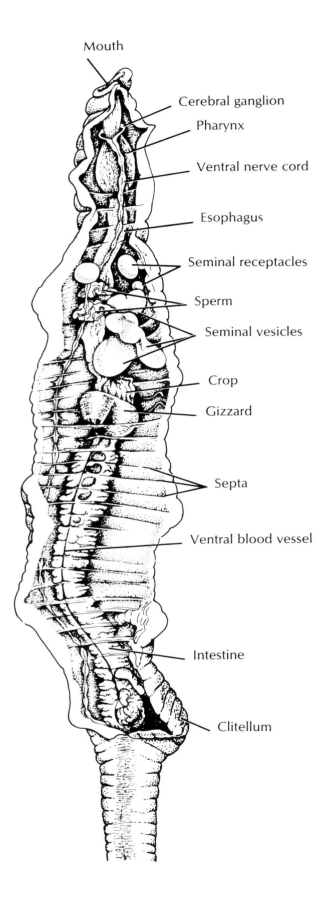

Mouth

Cerebral ganglion

Pharynx

Ventral nerve cord

Esophagus

Seminal receptacles

Sperm

Seminal vesicles

Crop

Gizzard

Septa

Ventral blood vessel

Intestine

Clitellum

FIG. 195

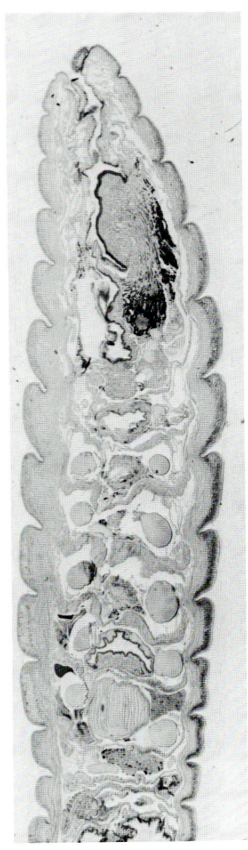

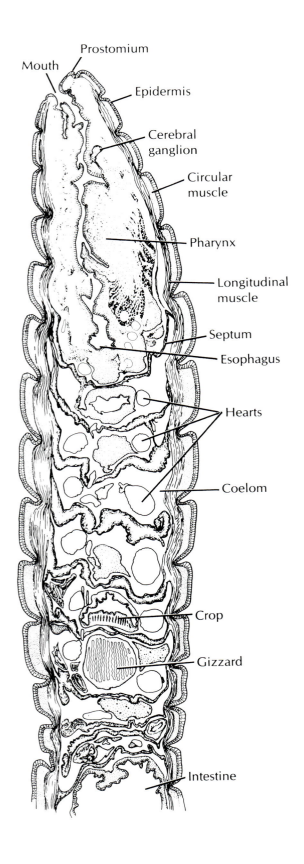

Mouth

Prostomium

Epidermis

Cerebral
ganglion

Circular
muscle

Pharynx

Longitudinal
muscle

Septum

Esophagus

Hearts

Coelom

Crop

Gizzard

Intestine

FIG. 196

FIG. 197. LM (×27) of a cross section through *Lumbricus terrestris* at the level of the intestine.

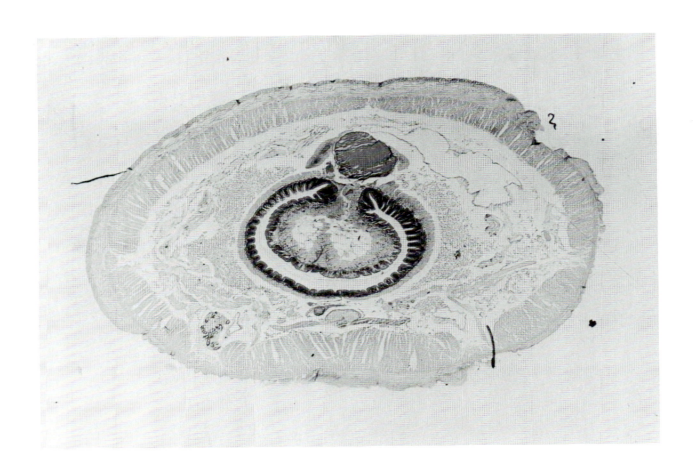

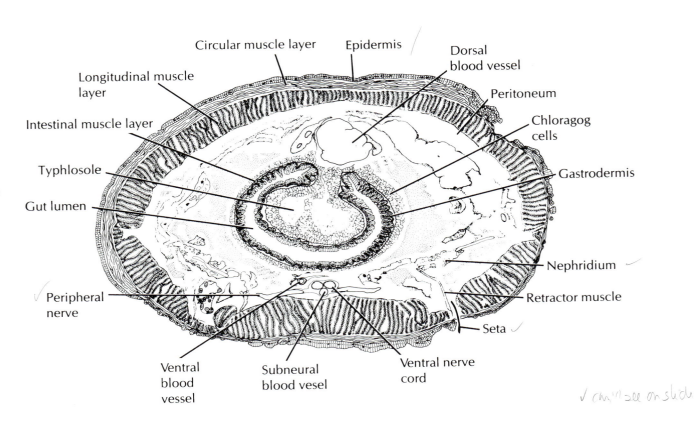

Circular muscle layer   Epidermis

Longitudinal muscle layer

Dorsal blood vessel

Peritoneum

Intestinal muscle layer

Chloragog cells

Typhlosole

Gastrodermis

Gut lumen

Nephridium

Peripheral nerve

Retractor muscle

Seta

Ventral blood vessel

Subneural blood vesel

Ventral nerve cord

FIG. 197

FIG. 198. The oligochaetes are hermaphroditic. During copulation there is a mutual transfer of sperm. When most species copulate, the male gonopore of one worm is directly opposite the spermathecal opening of the other worm, but not in the Lumbricidae. In this family, sperm transfer occurs externally, along temporary grooves formed by muscle contractions on the ventral body surface. After copulation, mucus is secreted around the clitellum and the anterior end of the worm. The clitellum then secretes a chitinlike covering, which becomes a cocoon. The cocoon is then slipped over the anterior end of the worm. The female deposits eggs into the cocoon as it passes the gonopores, and then sperm from the copulation are deposited into the cocoon as it passes the spermathecal opening. The ends of the cocoon constrict as they pass off the anterior end of the worm. The cocoons shown here are from *Eisenia foetida,* famous for living in manure piles. An egg is visible in one cocoon and a developing worm in another. (*Photograph by H. J. Linder, University of Maryland.*)

FIGS. 199 AND 200. Freshwater and marine oligochaetes are common, although they are less well known than their terrestrial counterparts. They are usually smaller than terrestrial worms and their external morphologies are more diverse. For example, *Dero vagus,* shown here in SEMs, has ciliated tufts scattered along its body. The posterior end (Fig. 199, ×540) has ciliated projections that serve as gills. The setae (Fig. 200, ×660) occur in clusters on each segment, generally dorsolaterally and ventrolaterally, and in the usual parapodial positions. Other parapodial structures are entirely absent.

164

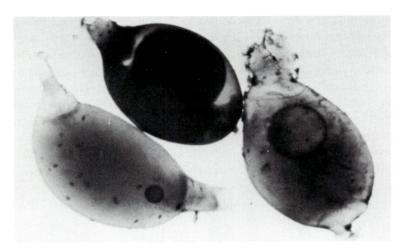

FIG. 198

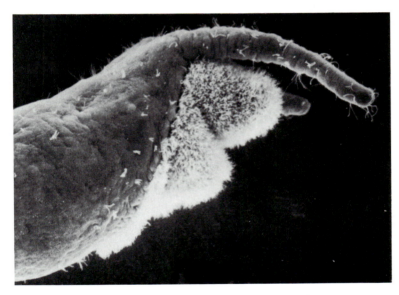

FIG. 199

FIG. 200

FIG. 201. Dissection of the medicinal leech *Hirudo medicinalis*. A typical leech has thirty-four segments, which have external annulations. The body is dorsoventrally flattened and has anterior and posterior suckers. Only one species has setae. There are no internal septa, and the coelom is filled with connective tissue called botryoidal tissue, which is composed of pigmented cells. Leeches are hermaphroditic, with the gonads restricted to a few segments. Cocoons are secreted by a clitellum that may not become obvious unless the leech is reproducing.

About 75 percent of leech species are bloodsuckers. Many of these species confine their feeding to one class of vertebrates. *Hirudo*, which prefers mammals, will feed on reptiles and amphibians as well. Bloodsucking leeches feed infrequently, but they usually ingest several times their own body weight when feeding occurs. The consumed blood cells are digested very slowly. *Hirudo* may require 200 days to digest a single meal. *(Dissection by E. Elliot, University of Maryland.)*

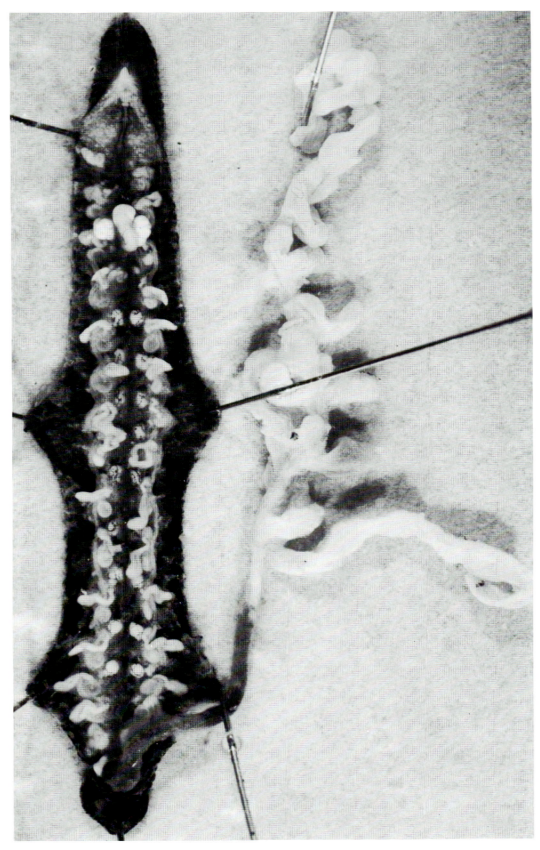

FIG. 201

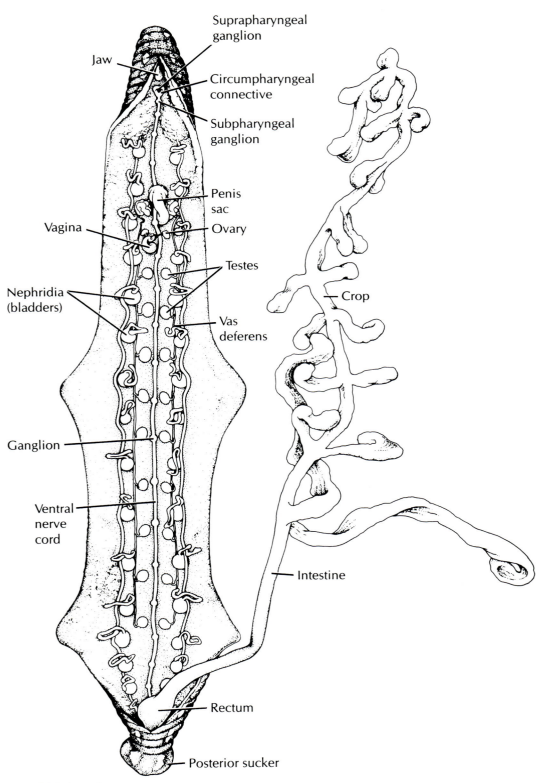

Suprapharyngeal
ganglion

Jaw

Circumpharyngeal
connective

Subpharyngeal
ganglion

Penis
sac

Vagina

Ovary

Testes

Nephridia
(bladders)

Vas
deferens

Crop

Ganglion

Ventral
nerve
cord

Intestine

Rectum

Posterior sucker

FIG. 201 (cont.)

FIG. 202. Close-up of anterior end of *Hirudo*.

FIG. 203. Close-up of midregion of *Hirudo*.

FIG. 204. Dorsal view (Fig. 204a) and ventral view (Fig. 204b) of *Hirudo*.

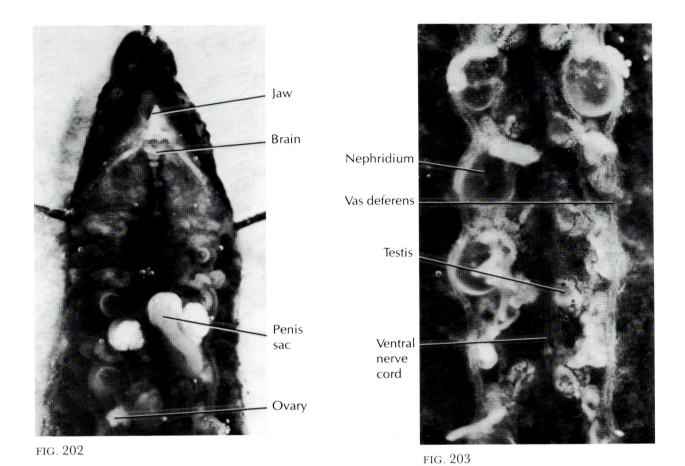

FIG. 202

FIG. 203

Jaw

Brain

Penis
sac

Ovary

Nephridium

Vas deferens

Testis

Ventral
nerve
cord

FIG. 204a

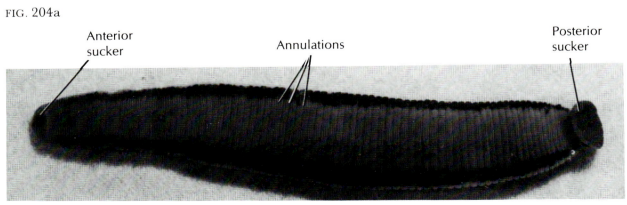

Anterior
sucker

Annulations

Posterior
sucker

FIG. 204b

SECTION VIII
Taxonomic Summary

Phylum Mollusca
  CLASS POLYPLACOPHORA
    Order Chitonida
      Chaetopleura apiculata
  CLASS GASTROPODA
   SUBCLASS PROSOBRANCHIA
    Order Mesogastropoda
      Littorina littorea
    Order Neogastropoda
      Busycon canaliculatum
      Thais lapillus
      Ilyanassa obsoleta
   SUBCLASS OPISTHOBRANCHIA
    Order Ascoglossa
      Elysia chlorotica
   SUBCLASS PULMONATA
      Limax maximus
  CLASS BIVALVIA
   SUBCLASS PALAEOTAXODONTA
    Order Nuculoida
      Yoldia limatula
   SUBCLASS CRYPTODONTA
    Order Solemyoida
      Solemya velum
   SUBCLASS PTERIOMORPHIA
    Order Mytiloida
      Mytilus edulis
      Aequipecten irradians
      Crassostrea virginica
   SUBCLASS HETERODONTA
    Order Veneroida
      Spisula solidissima
    Order Myoida
      Mya arenaria
  CLASS CEPHALOPODA
   SUBCLASS COLEOIDA
    Order Teuthoidea
      Loligo pealei

FIG. 205. The chitons are dorsoventrally flattened molluscs with eight overlapping shell plates. The plates, shown in this photograph of the dorsal surface of *Chaetopleura apiculata,* are of similar size and shape except for the first, the cephalic plate, and the last, the anal plate. The lateral edges of the plates are covered by the mantle, which secretes them. The mantle covers the plates to varying degrees in other chiton species. For example, the plates of the large gum boot chiton *Cryptochiton stelleri* are entirely covered. The lateral edge of the mantle, the girdle, is an extremely tough tissue in most species. It may be covered with scales, spicules (as in *Chaetopleura),* or bristles, or it may be smooth.

FIG. 206. The ventral surface of chitons is adapted to tightly adhere to rocky surfaces. The broad, flat foot and the girdle are used like a suction cup to form a slight vacuum which enables the chiton to grip the substrate. The mantle cavity of chitons is a trough, called the pallial groove, that borders the ventrolateral margin between the girdle and the foot. Many gills, the number of which varies with the size of the individual, lie in the groove. The gills are ciliated, as is usually the case in the molluscs (the cephalopods are a major exception). The cilia create a respiratory water current over the gill filaments. Water is drawn into the pallial groove from the anterior end of the animal, as well as laterally when portions of the girdle are elevated from the substrate. The water current leaves the pallial groove at the posterior end.

The serial repetition of body parts in chitons, that is, of shell plates and gills, as well as some of their larval characteristics have been interpreted to be remnant annelid characteristics and evidence of close phylogenetic ties between molluscs and annelid worms. The discovery of the monoplacophoran mollusc, *Neopilina galatheae,* which appeared to have internal metamerism, strengthened this view for a time. However, subsequent analysis of chiton (and *Neopilina*) morphology indicated that, while there is no doubt about the larval similarities between the chitons and polychaetes, the serial repetition of parts in adult chitons is not at all annelidlike. The shell plates appear on the surface of an unsegmented larva rather than from a budding zone, and the gills are added in a haphazard manner as the chiton grows (chitons often have unequal numbers of gills on each side). There is no indication of any internal metamerism at any stage. In fact, the loosely organized nervous system of the chiton is suggestive of flatworm affinities to some authors.

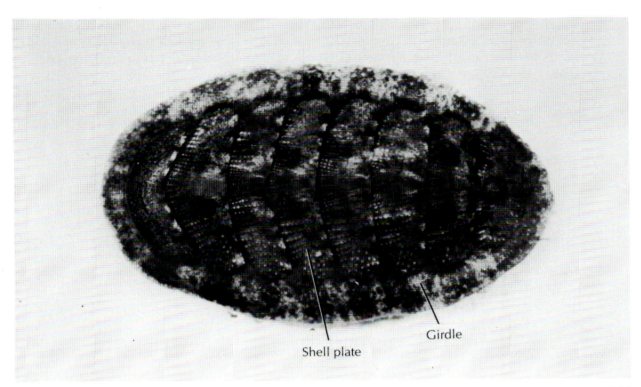

Girdle

Shell plate

FIG. 205

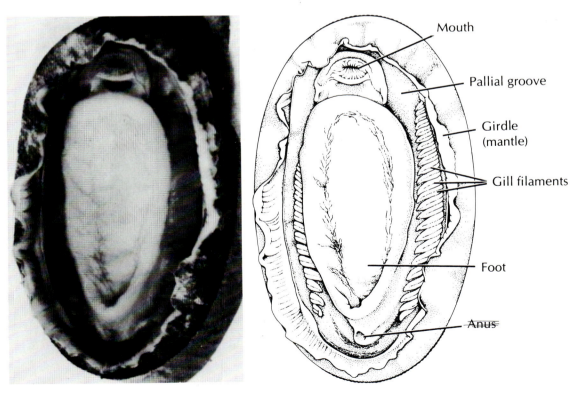

Mouth

Pallial groove

Girdle
(mantle)

Gill filaments

Foot

Anus

FIG. 206

FIG. 207. There is a tremendous diversity in both color and form of the shells of proso-branch snails. However, the soft parts are similar among species. *Busycon canaliculatum*, shown here removed from its shell, is typical. The mantle cavity has been cut open over the head and pinned back.

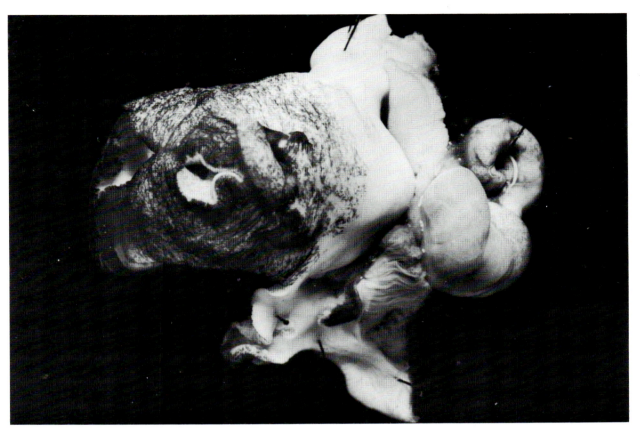

FIG. 207

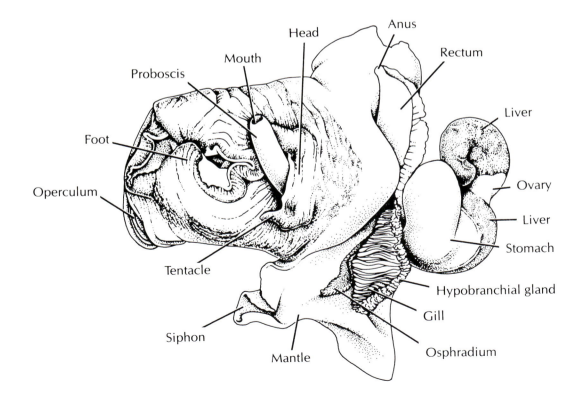

FIG. 208. Male specimen of *Busycon*. In this species, fertilization is internal, and sperm are transferred by means of a large penis that lies on the right side of the head. *Busycon* eggs are protected by an egg case secreted by the nidamental gland of the female. This species has no free-living larval stages. Miniature adults hatch from the egg case.

FIG. 209. Ventral view of *Busycon*. The operculum is a protective structure that acts as a door. It seals off the soft parts of the animal when they are retracted into the shell by the columellar muscle, which attaches the snail to its shell.

FIG. 210. The nervous system of many gastropod species shows a high degree of cephalization. Most of the ganglia are located in a ring surrounding the anterior portion of the esophagus. The *Busycon* brain, shown here with the esophagus removed and many of the axons cut, is firmly embedded in the tissue surrounding the esophagus. Pigmented cell bodies are visible, but they are small compared with the giant cell bodies of neurons in some of the opisthobranch snails *(Aplysia, Navanax, Elysia)*. The nervous system of gastropods is much more organized than that of chitons.

FIG. 211. The feeding organ of all gastropods is a toothed strap called the radula located immediately inside the mouth and surrounded by the protrusible proboscis. The proboscis of *Busycon* has been slit open in this photograph to reveal the radula (see also Figs. 212–215). The radula and its underlying support, the odontophore, are operated by a complex set of muscles. During feeding, the radula is extended to the substrate and pulled back and forth in a rasping motion by the muscles. The teeth point posteriorly; thus they are effective on the backward stroke. In some species the radula is modified to slit or pierce algae or prey. Some species have poison glands associated with the radula. Most infamous of these are the cone shells *(Conus)*, which have a single, hollow radular tooth that is thrust into the prey with a neurotoxin. The toxin of some *Conus* species has killed more than one unwary shell collector.

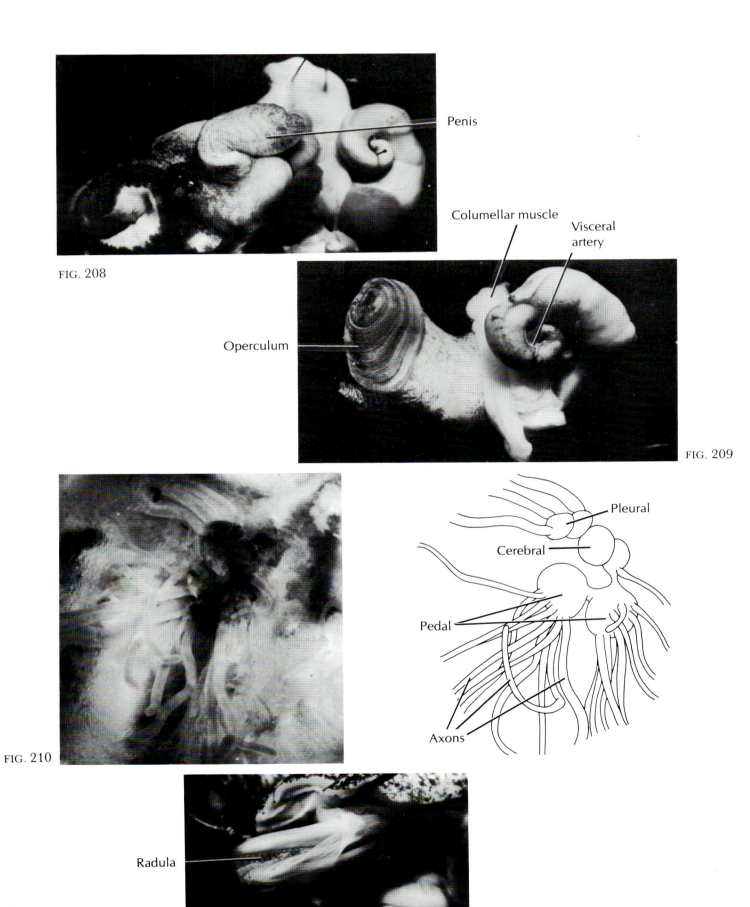

Penis

FIG. 208

Columellar muscle

Visceral artery

Operculum

FIG. 209

Pleural

Cerebral

Pedal

Axons

FIG. 210

Radula

FIG. 211

FIG. 212–215. The morphology of the radular teeth in snails varies from species to species. Presumably, they are appropriately designed for efficient rasping of the type of food preferred by each species. The radulas shown in these SEMs are from prosobranch snails that have very different feeding habits. *Busycon canaliculatum* (Fig. 212, ×50) feeds on bivalves but does not bore the shells; instead it thrusts the proboscis between the valves. *Ilyanassa obsoleta* (Fig. 213, ×170) is a scavanger that feeds both on organic material in the sediment and on carrion. *Thais lapillus* (Fig. 214, ×420) drills into the shells of bivalves, snails, and barnacles using both radular action and chemical digestion. *Littorina littorea* (Fig. 215, ×330) is an herbivore.

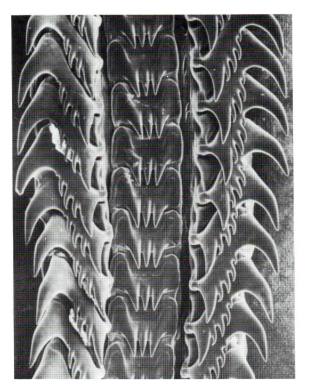

FIG. 212

FIG. 213

FIG. 214

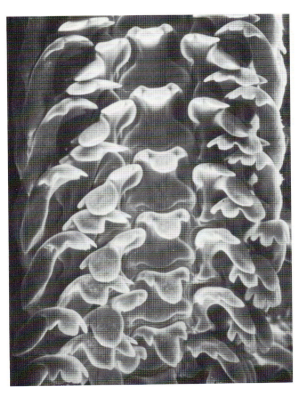

FIG. 215

FIG. 216. The gastropods, unlike the rest of the molluscs, are widely represented in the terrestrial fauna. Terrestrial snails have no gills; the mantle cavity has become a highly vascular lung. The lung, which accounts for the name of the subclass, Pulmonata, opens to the outside through the pneumostome, shown patent here in the slug *Limax maximus* (Fig. 216c). The pneumostome is surrounded by a sphincter that can close it tightly. The lung is ventilated by the alternate arching and flattening of the floor of the mantle cavity.

Many pulmonates have shells, but the group also includes the land slugs, which lack shells, such as *Limax maximus* shown here in dorsal dissection (Fig. 216a and b). Pulmonates are also common in freshwater, and there are a few intertidal marine species. Some of the aquatic forms have a secondary gill in the mantle cavity (pseudobranch). Some no longer breath air at all; the lung fills with water.

The pulmonates are hermaphroditic. Copulation is a spectacular event in *Limax*. Two animals climb a tree and, after courtship, they wind around each other in a spiral and then dive head first into the air suspended by a mucous cord. The animals copulate while suspended and afterward climb up the mucous cord, eating it as they go. *(Dissection by L. H. Smith, University of Maryland.)*

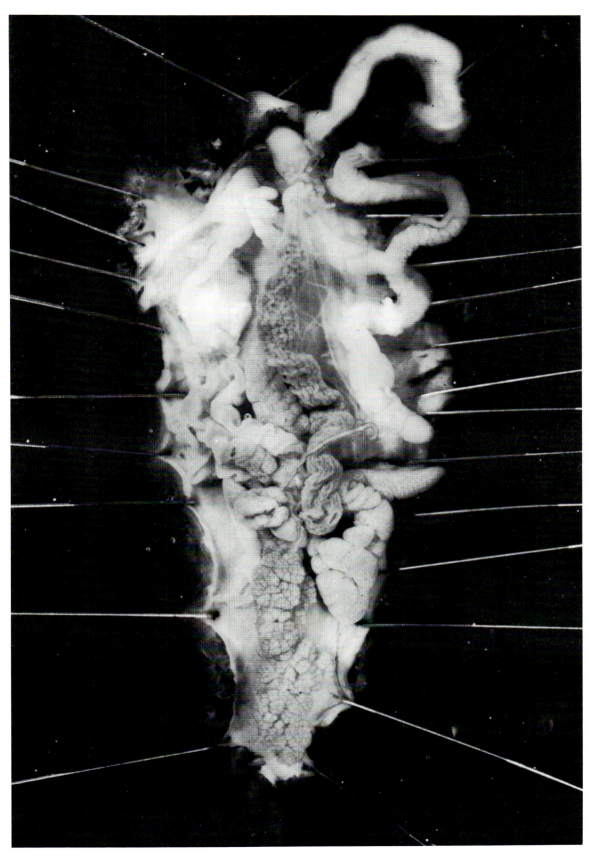

FIG. 216a

FIG. 217. Gastropods, like the polychaetes and chitons, have a trochophore larval stage. In the snails, the trochophore develops into a second larval stage, the veliger. This stage is covered by a coiled shell and swims by means of a bilobed, ciliated structure called the velum. This LM shows a *Busycon* veliger that has been removed from the egg case.

FIG. 218. SEM ($\times$600) of a 5-day-old veliger of the opisthobranch slug *Elysia chlorotica*. The sensory structures between the velar lobes are clearly seen.

FIG. 219. There are also trochophore and veliger larval stages in the bivalves. The veliger has a bivalve shell, so the larva has the appearance of a tiny clam. The velum is protruded through the gape in the shell and, as in the snail, is a highly ciliated organ used for swimming. This SEM ($\times$300) shows a *Crassostrea virginica* veliger, 16 to 18 days old, about ready to set. The veligers of both snails and bivalves can retract the velum entirely into the shell and either seal the shell off with an operculum (gastropods) or tightly close the valves (bivalves). Thus protected, both types of veligers can pass unscathed through the digestive systems of many plankton feeders. *(Photograph by D. B. Bonar, University of Maryland.)*

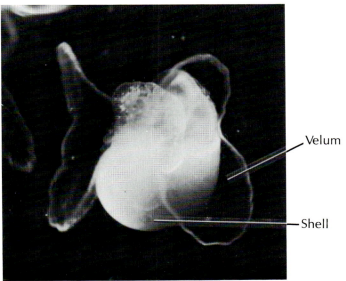

FIG. 217

Velum

Shell

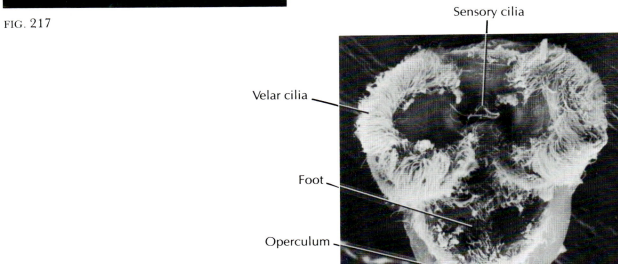

Sensory cilia

Velar cilia

Foot

Operculum

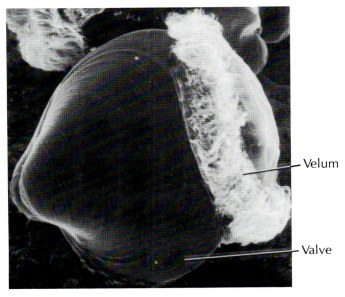

FIG. 219

Velum

Valve

FIG. 220. In the bivalve molluscs, as in the gastropods, there is tremendous diversity in shell shape, ornamentation, and color. The morphology of the soft parts is more variable in bivalves than in gastropods, but the anatomy of a standard clam enables the interpretation of the anatomy of other bivalve species. The lamellibranch clam shown dissected here, the surf clam *Spisula solidissima,* displays the standard anatomy.

Clams have no head, but the end where the mouth is located, between the palps, is considered anterior. The sense organs are usually not as well developed in bivalves as in gastropods (but see Fig. 224) and the nervous system is less complex. Most species are filter feeders and rely on complexly ciliated gills to create both an inward respiratory water current laden with plankton and an exhalant current containing wastes, and to filter food (see also Figs. 230–246).

The typical clam has a symmetrically clam-shaped shell with a springy, proteinaceous hinge at the dorsal edge, between the umbos, which are the oldest parts of the valves. In *Spisula,* the hinge is opposed by a pair of equal-sized adductor muscles (the isomyarian condition) that close the shell tightly when they contract. The shell is secreted and maintained by the mantle, as is usual in the molluscs. The outer surface of the valve is usually covered with a periostracum, which is continuous with the outer mantle edge. Locomotion is accomplished by means of the foot, which is such a prominant part of the anatomy that it was the basis in older taxonomic systems for the name of the class: Pelecypoda, or hatchet foot.

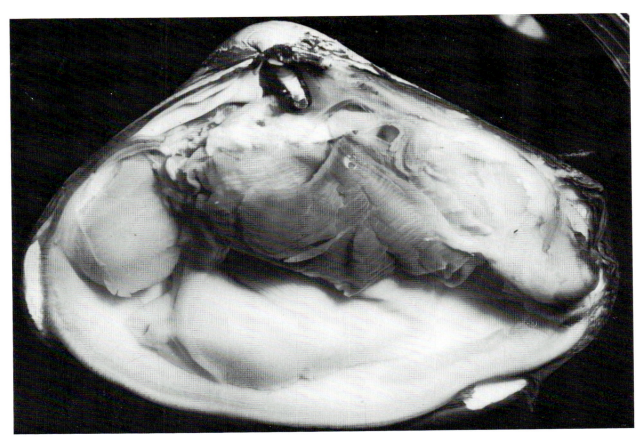

FIG. 220

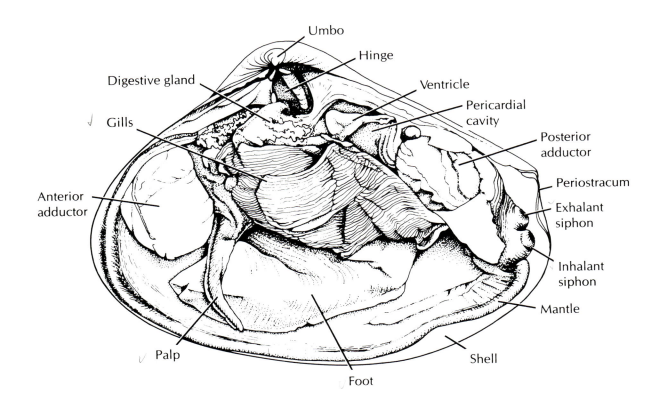

Umbo

Hinge

Digestive gland

Ventricle

Pericardial cavity

√ Gills

Posterior adductor

Anterior adductor

Periostracum

Exhalant siphon

Inhalant siphon

Mantle

√ Palp

Shell

Foot

FIG. 221. *Spisula* specimen of Fig. 220 with the ctenidia (gill) cut away to expose the underlying viscera, which consists mostly of foot and digestive gland in this species. The palps that originate around the mouth are also revealed by removal of the overlying ctenidia.

FIG. 222. Close-up of the pericardial region of *Spisula* with the pericaridal membrane removed. The single ventricle is a muscular structure anchored anteriorly to the aorta, laterally to the pair of thinner-walled auricles, and posteriorly to the rectum. Interestingly, the rectum traverses the ventricular lumen in this and many other species of bivalves. Aside from being the source of some bad jokes, the reason for this association between ventricle and rectum is obscure. The bivalve circulatory system is a low-pressure, open system. The heart produces blood pressures of no more than a few millimeters of mercury. Thus the blood flow is very sluggish and often, especially in areas of high flow resistance such as gills, accessory hearts or contractile blood vessels occur. The pericardial cavity is the only remnant of the coelom. Hydrostatic skeletal functions have been assumed by the blood sinuses.

FIG. 223. The bivalve nervous system is anatomically simpler than that of the snails. The bivalves have three paired ganglia: the cerebropleural ganglion and the visceral ganglion located just ventral to the anterior and posterior adductor muscles, respectively (in isomyarian species), and the pedal ganglion located in the foot. Paired nerve cords arise from the cerebropleural ganglion and extend to both the visceral and pedal ganglia. A dissected visceral ganglion from *Mya arenaria* is shown here. The large nerves are the cerebropleural–visceral connectives. The smaller nerves leave the ganglion to innervate, perhaps, the posterior adductor muscle or the siphons. *(Dissection by L. Beres, University of Maryland.)*

FIG. 224. The sedentary bivalves have fairly simple sense organs. A pair of statocysts is usually associated with the pedal ganglion. Chemosensitive and tactile cells are associated with the mantle edge or pallial tentacles. Simple eyespots, called pigment cup ocelli, are often located on the edges of the mantle or siphons. In the swimming bivalves, the eyes are often well developed. The eyes of the blue-eyed scallop, *Aequipecten irradians,* shown here (see also Fig. 228), contain a cornea, a lens, and a retina.

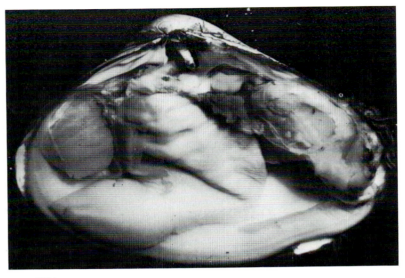

FIG. 221

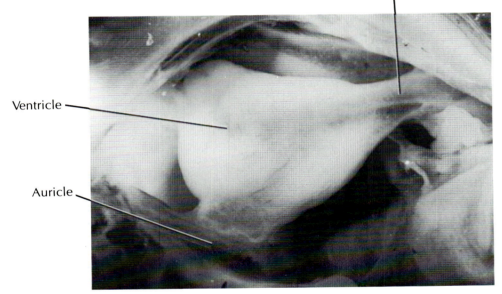

Rectum

Ventricle

Auricle

FIG. 222

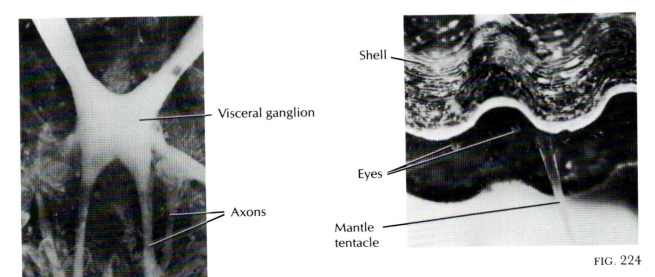

Visceral ganglion

Axons

FIG. 223

Shell

Eyes

Mantle
tentacle

FIG. 224

FIG. 225. Clams in subclass Palaeotaxodonta are considered to be relatively primitive. Most are deposit feeders and thus are very active burrowers with well-developed feet. Since the gills are not feeding structures in this group, they are not as elaborate as in the Pteriomorphia or Heterodonta (see Figs. 230–233). The gill type is termed protobranch. *Yoldia limatula,* shown in dissection here, is a very common example of a protobranch bivalve. The palaeotaxodonts are also characterized by the type of hinge teeth evident on the dorsal edge of the *Yoldia* valve. The palp appendages emerge between the valves and feel the substrate for suitable food particles. The particles are carried along the palp appendages in a ciliated groove to the labial palps, where ciliary sorting takes place. Then, eventually, the appropriately sized particles are moved to the mouth along additional ciliated channels.

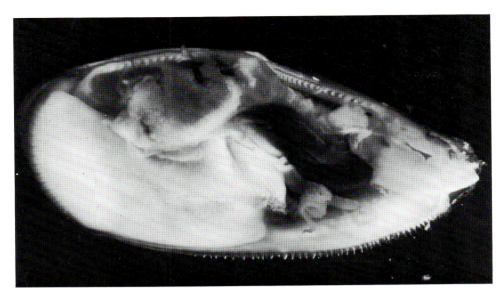

FIG. 225

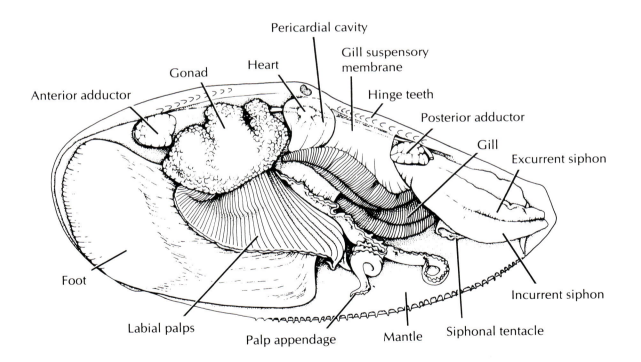

Anterior adductor

Gonad

Heart

Pericardial cavity

Gill suspensory membrane

Hinge teeth

Posterior adductor

Gill

Excurrent siphon

Foot

Labial palps

Palp appendage

Mantle

Siphonal tentacle

Incurrent siphon

FIG. 226. *Mytilus edulis,* shown here in half-shell dissection, is a common anisomyarian bivalve, that is, having unequally sized adductor muscles. The anterior adductor is reduced in size relative to the posterior adductor. *Mytilus* has a well-developed filibranch gill and is exclusively a filter feeder.

Mytilus attaches to rocks, wharf pilings, oysters, and other solid substrates by means of tough, proteinaceous threads called byssal fibers. These are secreted as a liquid by a byssal gland located near the base of the foot. The liquid flows along a groove in the foot and to the substrate. When the protein solidifies, the foot is removed and a byssal thread is left.

FIG. 227. The byssus, a group of byssal threads, is attached to muscles, the anterior and posterior byssus retractor muscles, by way of the byssal gland. These muscles are clearly visible in the dissection of *Mytilus* shown here in which the surrounding viscera have been removed. Contraction of the byssus retractor muscles pulls the mussel tightly to the substrate.

The anterior byssus retractor muscle has been studied extensively by muscle physiologists because it (along with bivalve adductor muscles) exhibits a property called catch. The muscle contracts and stays contracted for very long periods of time (oyster adductors can remain contracted for many days) without any nervous stimulation after the initial stimuli and with little metabolic expenditure, that is, the muscle contracts and then "catches" in the contraction state. The physiological basis of the catch phenomenon is unclear, but it seems to be related to the presence of a special contractile protein called paramyosin that is present in the muscle in addition to the usual actin and myosin. This type of muscle, which allows the bivalves to "clam up" for lengthy periods, has been found only in the bivalves and in fresh water horse-hair worms (phylum Nematomorpha). The function of the catch muscle in the worms, where it is found in the body wall, is entirely unknown.

FIG. 226

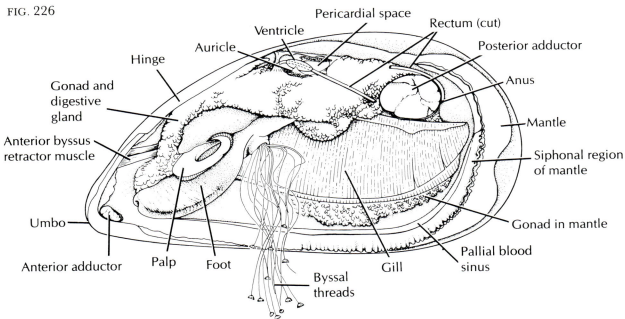

Hinge

Gonad and
digestive
gland

Anterior byssus
retractor muscle

Umbo

Anterior adductor

Palp

Foot

Byssal
threads

Auricle

Ventricle

Pericardial space

Rectum (cut)

Posterior adductor

Anus

Mantle

Siphonal region
of mantle

Gonad in mantle

Pallial blood
sinus

Gill

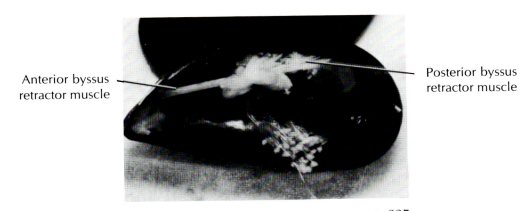

Anterior byssus
retractor muscle

Posterior byssus
retractor muscle

FIG. 227

FIG. 228. Scallops, such as the *Aequipecten irradians* specimen dissected here, are unattached, bottom dwellers. When not swimming, scallops lie on one valve, and, in many species, that valve is flattened. The foot is small and the anterior adductor is absent. Thus, these bivalves are monomyarian. *Aequipecten* swims by rapidly opening and closing the valves, using contractions of the adductor muscle opposed by a very springy hinge. As the valves are snapped shut, a jet of water is expelled from between the mantle margins. A muscular fold of the mantle, called the velar fold, controls the direction of the jet propulsion. Unlike most bivalves, scallops are hermaphroditic. They also have well-developed sensory structures (eyes and tentacles; see also Fig. 224) along the mantle margins.

FIG. 228

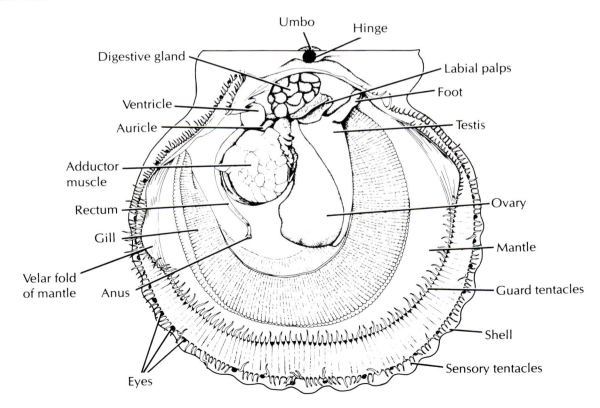

FIG. 229. Oysters are a sessile form of bottom-dwelling bivalve. They are usually attached in place by one valve, which becomes cemented to the substrate by a calcareous secretion as the oyster grows. Oyster anatomy, shown in this dissection of *Crassostrea virginica,* is different from that of the standard clam. As in the scallop, the anterior adductor of the oyster is missing entirely. The foot is also absent. Some oyster species are protandric hermaphrodites. In *Crassostrea,* the sexes are separate but unstable; individuals commonly change sex after spawning.

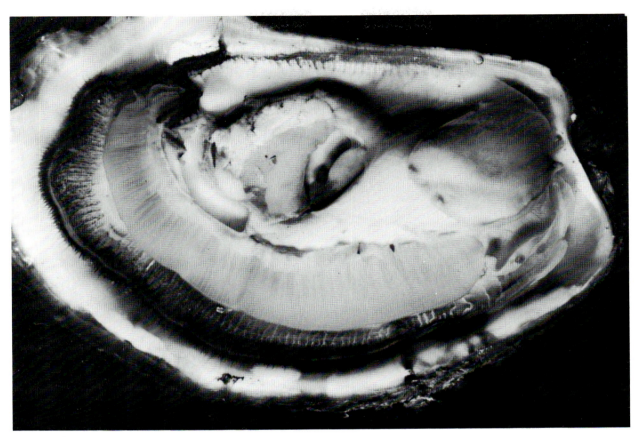

FIG. 229

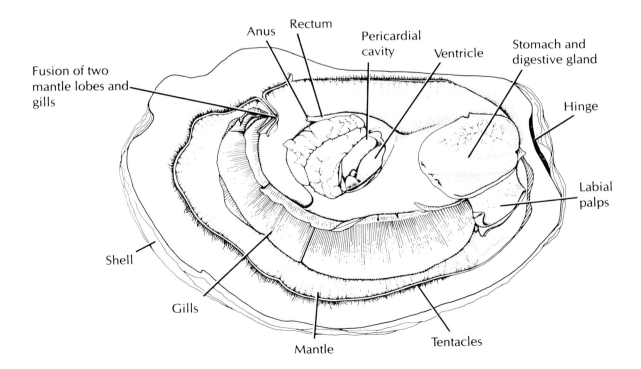

Fusion of two mantle lobes and gills

Anus

Rectum

Pericardial cavity

Ventricle

Stomach and digestive gland

Hinge

Labial palps

Shell

Gills

Mantle

Tentacles

FIGS. 230 AND 231. The protobranch gill is the simplest type of bivalve gill; it is not a feeding structure. Each gill filament is highly ciliated as evident in the SEM of a short length of *Yoldia limatula* gill (Fig. 230, ×250). The cilia at the outer edges of the filaments create the respiratory water current. There are no cilia between filaments (Fig. 231, × 660) except in regions where ciliated tufts interlock with adjacent filaments.

FIG. 232. SEM (×1760) showing the interlocking cilia on adjacent filaments in a *Solemya velum* gill. *Solemya* is a cryptodont protobranch.

FIG. 233. Diagram of a typical protobranch gill.

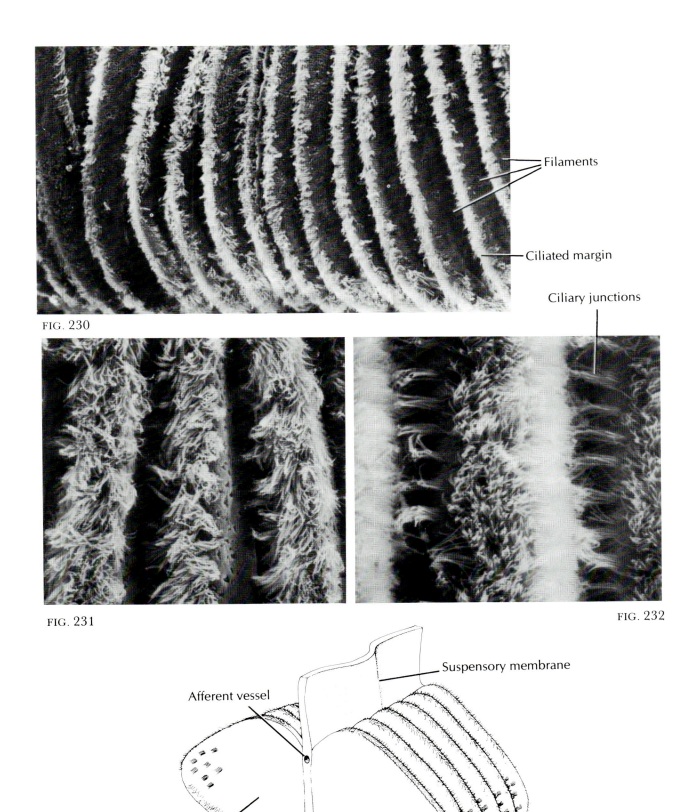

FIG. 230

Filaments

Ciliated margin

Ciliary junctions

FIG. 231

FIG. 232

Suspensory membrane

Afferent vessel

Filament

Efferent vessel

Ciliary junctions

FIG. 233

FIGS. 234 AND 235. SEMs of the outer face of the filibranch gill of *Mytilus edulis,* on which there are more types of cilia than on the protobranch gill. The lateral cilia between adjacent filaments create the water current. At intervals along the filament, the lateral cilia of adjacent filaments interlock (Fig. 234, ×220) and hold the entire structure together and the filaments in register. At higher magnification (Fig. 235, ×1210), the lateral cilia and ciliary junctions between adjacent filaments are clearly seen. Distal to the lateral cilia are longer cilia that do not beat, the laterofrontal cilia. They are the filter portion of the gill filament.

The water current created by the lateral cilia is drawn through the laterofrontal mesh, which filters out particles. The particles are transferred in a mucous sheet to the frontal cilia on the distal face of the gill filament. The frontal cilia (there may be more than one type) sort the particles by size and conduct the appropriately sized ones, still trapped in mucus, toward the ciliated food groove. The location of the food groove varies from species to species, but it always runs longitudinally along the gill. The cilia in the groove move the food toward the mouth. Additional ciliary sorting of food particles occurs in the food groove. Unsuitable particles are dropped off into the mantle cavity. Just before the food particles reach the mouth, they are sorted one last time by the cilia of the labial palps.

FIG. 236. Diagram of a typical filibranch gill.

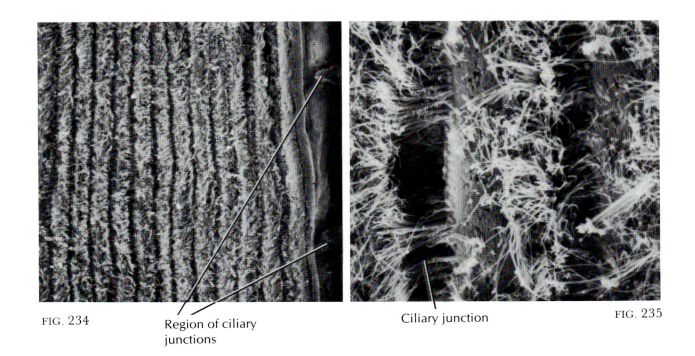

FIG. 234

Region of ciliary
junctions

Ciliary junction

FIG. 235

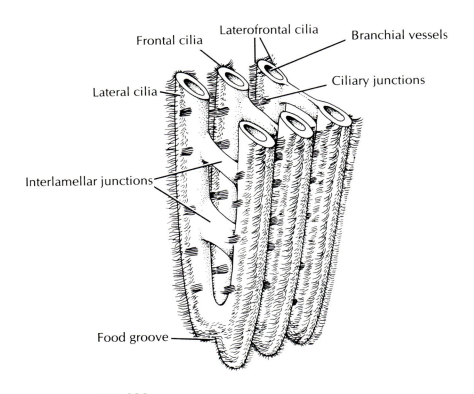

Frontal cilia

Laterofrontal cilia

Branchial vessels

Lateral cilia

Ciliary junctions

Interlamellar junctions

Food groove

FIG. 236

FIGS. 237–239. The filibranch gill of the scallop *Aequipecten irradians* is slightly modified. Tiny flaps called cilifers (Fig. 237, ×130 and 238, ×250) bear the cilia that hold the filaments together. The cilia (called junctional or J-cilia) on the cilifers interlock in pairs consisting of one cilium from each of two adjacent cilifers (Fig. 239, ×9000). The arrows in Fig. 239 indicate where the tip of one J-cilium from a pair is hooked around its mate. (From C. Reed-Miller and M. J. Greenberg, "The ciliary junctions of scallop gills: The effects of cytochalasins and concanavalin A.," *Biol. Bull.* 163 [1982]:225–39, Figs. 1–3.)

Cilifer

FIG. 237

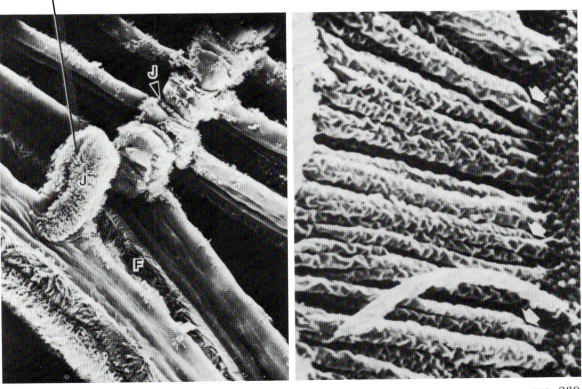

FIG. 238

FIG. 239

FIGS. 240 AND 241. The morphology of the oyster gill is peculiar. The gill filaments are longitudinally infolded, creating a plicate margin on the external surfaces. Each of the infoldings is called an ordinary filament (Fig. 240, SEM, ×200). Incurrent pores called ostia open between the ordinary filaments. The cilia on the ordinary filaments are similar to that of the filibranch gill. The plicate filaments are separated from each other by principal filaments that are not folded (Fig. 241, SEM, × 890). The longitudinal ciliary tract on the principal filament carries particles only in an upward direction. The filaments of the oyster gill are held together by tissue junctions between the filaments, a eulamellibranch characteristic (see Figs. 242–244). Because of this mixed morphology, the oyster gill, shown in these SEMs *(Crassostrea virginica)*, is called a pseudolamellibranch type.

FIG. 240

Ordinary filament

Principal filament

Frontal cilia

FIG. 241

Plicate ordinary filament

FIGS. 242 AND 243. The ciliature of the eulamellibranch gill is basically the same as that of the filibranch gill. However, instead of being held together by ciliary junctions, the filaments of the eulamellibranch gill are fused at their proximal bases into a continuous sheet. The gill is studded with pores, the ostia (Fig. 242, LM, ×65), that allow water to pass from the inhalant side to the exhalant side. The diameters of the ostia are controlled by the filament musculature, but the largest diameter corresponds closely to the diameter of the ova that must pass through the ostia during spawning. Both the ostia and the filaments of the eulamellibranch gill are supported by proteinaceous skeletal rods (Fig. 243, LM, ×420).

FIGS. 244 AND 245. SEMs of the eulamellibranch gill of *Spisula solidissima.* The food groove is clearly visible at the bottom edge (Fig. 244 ×210). At higher magnification (Fig. 245, ×880), the dense ciliation of the gill obscures the view of the underlying structure entirely. The blebbed tips of the cilia are probably an artefact of fixation.

FIG. 246. Diagram of a eulamellibranch gill.

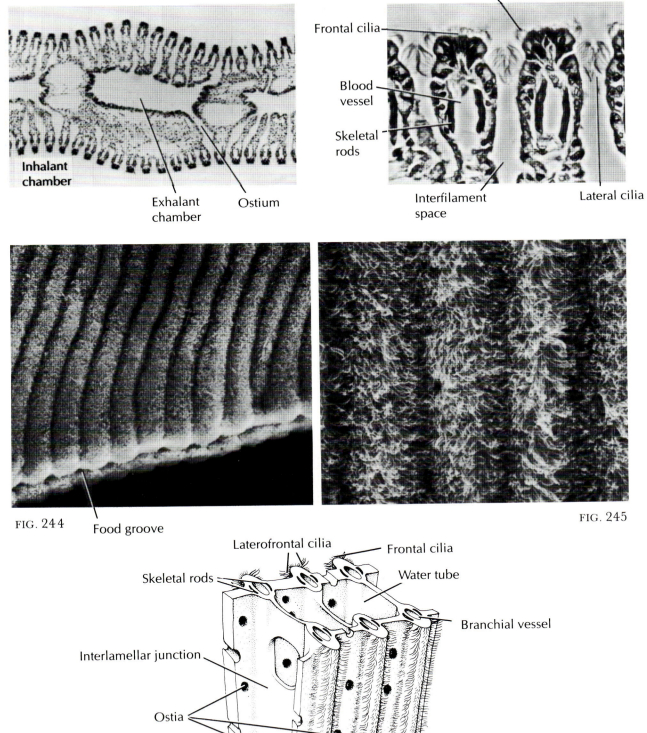

FIG. 242

Inhalant chamber

Exhalant chamber   Ostium

Laterofrontal cilia   FIG. 243

Frontal cilia

Blood vessel

Skeletal rods

Interfilament space   Lateral cilia

FIG. 244   Food groove

FIG. 245

Laterofrontal cilia   Frontal cilia

Skeletal rods   Water tube

Branchial vessel

Interlamellar junction

Ostia

Lateral cilia

FIG. 246

FIG. 247. The cephalopod molluscs have been called the invertebrate zenith by some authors with considerable justification. The squids and octopods are relatively intelligent animals, compared, for example, with clams or sea worms, and they exhibit complex behavior patterns. The cephalopod nervous system is well developed. The eyes are excellent visual organs, very similar in construction to vertebrate eyes. Cephalopods have a closed, high-pressure circulatory system, and their gills lack cilia. They are fast, jet-propelled swimmers (including the bottom-dwelling octopods) and are very active carnivores. *Loligo pealei,* shown here in dissection, is dioecious, as are the rest of the cephalopods. This specimen is male. The shell in cephalopods is often internalized (the pen in squids) or lacking (octopods). Of the extant cephalopods, only *Nautilus* has an external shell.

The giant squid *Architeuthis* occupies a permanent place in folklore. The giant axon and giant synapse of the smaller squids, such as *Loligo pealei,* dissected here, have a permanent place in neurobiology. A great deal of what is currently known about neuron function has been gained in the study of the unique cephalopod nervous system.

FIG. 248. Close-up of the reproductive organs and the respiratory and circulatory organs of a male *Loligo.*

FIG. 247

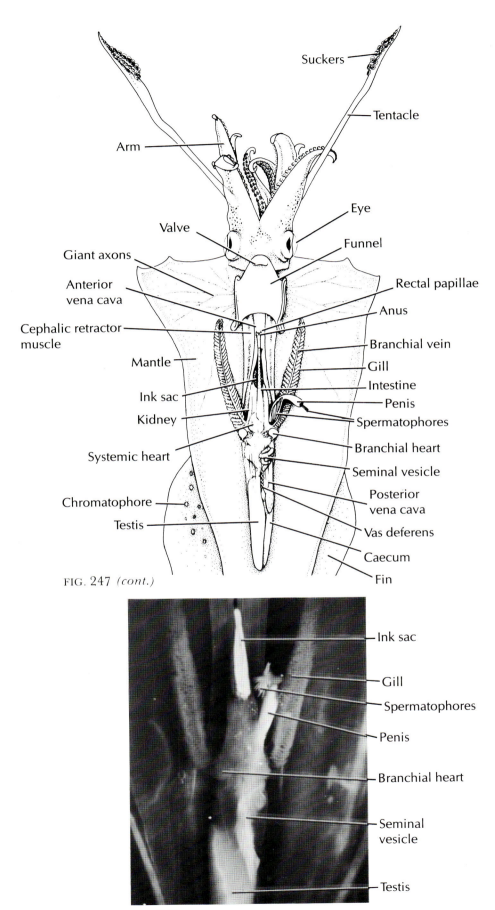

Suckers

Tentacle

Arm

Eye

Valve

Funnel

Giant axons

Rectal papillae

Anterior
vena cava

Anus

Cephalic retractor
muscle

Branchial vein

Gill

Mantle

Intestine

Ink sac

Penis

Kidney

Spermatophores

Systemic heart

Branchial heart

Seminal vesicle

Chromatophore

Posterior
vena cava

Testis

Vas deferens

Caecum

Fin

FIG. 247 *(cont.)*

Ink sac

Gill

Spermatophores

Penis

Branchial heart

Seminal
vesicle

Testis

FIG. 248

FIG. 249. In a gravid female squid, such as this *Loligo,* the gonad is filled with large, yolky eggs. During copulation, the female receives several spermatophores from the male, either directly into the mantle cavity or into a seminal receptacle. Immediately after fertilization, the eggs are deposited in large gelatinous masses.

Nidamental gland

Ovary

Oviductal
opening

FIG. 249

FIG. 250. The development of cephalopods is unmolluscan in that it is direct; there are neither trochophore nor veliger larvae. The eggs develop directly into juveniles, which hatch from the egg mass. The developing *Loligo pealei* embryo shown here in SEM (×65) was removed from the egg mass when it was 6 to 7 days old; it would have hatched in about one more week.

FIG. 251. Close-up of the head of *Loligo pealei,* dissected to show the brain and buccal mass. In the well-developed cephalopod nervous system, the usual molluscan ganglia are fused into a circumesophageal brain. The buccal mass forward of the brain contains powerful, parrot-beaked jaws surrounding a small radula. The beak is used to bite chunks of tissue from prey captured by the arms; the radula assists in ingestion.

FIG. 252. The giant axons originate from and the giant synapse is located in the stellate ganglion of *Loligo pealei,* shown in this dissection. The axons innervate the mantle muscle. Their large diameter (300 to 700 μm) facilitates the conduction of impulses so the squid can rapidly contract the mantle and force a propulsive jet of water out the funnel. This is an escape response. The large axon also allows the neurobiologist easy access to the inner side of the neuronal membrane.

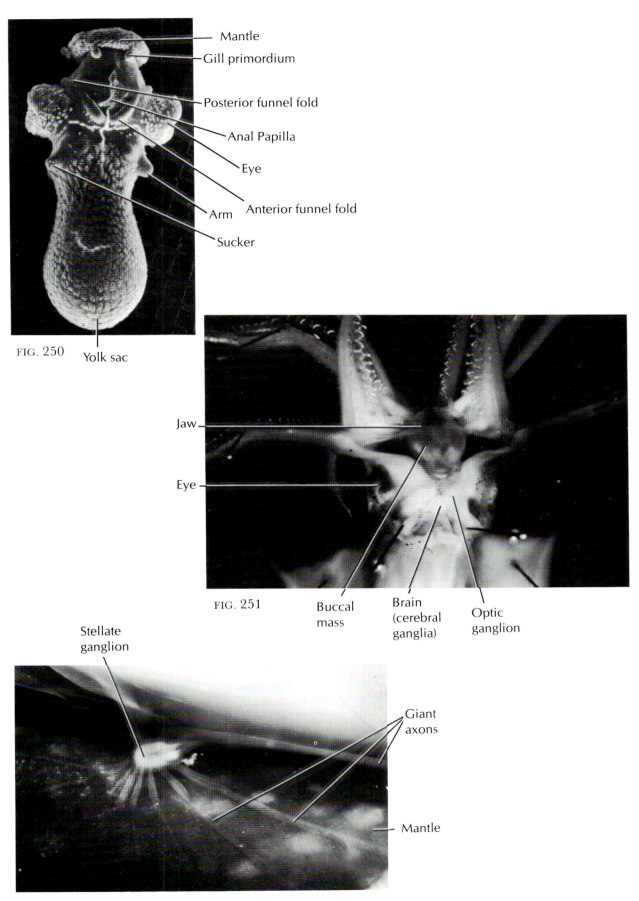

Mantle

Gill primordium

Posterior funnel fold

Anal Papilla

Eye

Anterior funnel fold

Arm

Sucker

FIG. 250   Yolk sac

Jaw

Eye

FIG. 251   Buccal mass   Brain (cerebral ganglia)   Optic ganglion

Stellate ganglion

Giant axons

Mantle

FIG. 252

SECTION IX
Taxonomic Summary

Phylum Arthropoda
  Subphylum Chelicerata
    CLASS MEROSTOMATA
     SUBCLASS XIPHOSURA
       Limulus polyphemus
    CLASS PYCNOGONIDA
      (Sea Spider, Species Unknown)
  Subphylum Crustacea
    CLASS BRANCHIOPODA
     Order Anostraca
       Eubranchipus
     Order Diplostraca
       Daphnia
    CLASS OSTRACODA
       Cypris
    CLASS COPEPODA
     Order Cyclopoida
       Cyclops
       Mesocyclops edax
    CLASS CIRRIPEDIA
       Lepas anatifera
    CLASS MALACOSTRACA
     Order Isopoda
       Oniscus asellus
     Order Amphipoda
      Suborder Gammaroidea
       Gammarus
       Jassa
      Suborder Caprellidea
       Caprella geometrica
     Order Decapoda
      Suborder Pleocyemata
       Infraorder Actactidea (Crayfish)
       Infraorder Anomura
        Emerita talpoida
       Infraorder Brachyura
        Cancer irroratus
        Cancer borealis
        Callinectes sapidus

FIGS. 253 AND 254. The horseshoe crab, *Limulus polyphemus,* is one of the few species of the class Merostomata still extant. Most merostomatid species are extinct. *Limulus* is a large and strange-looking animal. For a time it was thought that *Limulus* might have originated from among the arachnids. Another theory placed *Limulus* close to the base of the vertebrate phylogenetic tree. The larva of *Limulus* superficially resembles the extinct trilobites (see Fig. 262), so attempts have been made to ally the two groups phylogenetically.

The *Limulus* specimen shown here is female. Externally, the body has two main sections. The anterior section, which bears the lateral eyes dorsally and the legs ventrally, is called the prosoma or cephalothorax. The posterior part of the body, which bears the gills and spike-like tail and is hinged to the prosoma, is called the opisthosoma or abdomen (Fig. 253). The body is covered externally by a strong exoskeleton.

On the ventral side of the prosoma (Fig. 254) are seven pairs of appendages, corresponding to seven body segments. The three-segment chelicerae (from which the subphylum takes its name) make up the first pair of appendages, just anterior to the mouth. The next five pairs are walking legs aligned along the central mouth. The distal segment of the first pair of walking legs in mature males is modified to clasp onto the abdomen of the female during fertilization (see Fig. 255). The middle four pairs of walking legs are identical; they have six segments, with a chelate segment distally and a masticatory segment, called the gnathobase, proximally. The fifth pair of walking legs is not chelate; instead of a claw, the distal end comprises one elongated and four flattened spines used for pushing and digging. The seventh pair of prosomal appendages are blunt, unsegmented spines called chilaria.

There are six sets of flaps on the ventral side of the abdomen. These flaps may represent the lateral fusion of paired abdominal appendages. The first flap, called the genital operculum, protects the gills and contains the genital openings (see also Fig. 257). The remaining five flaps bear stacks of gill lamellae proximally, called book gills (see also Fig. 257), which are protected distally by gill opercula. The gill flaps beat synchronously to aerate the gills as the animal plows through the mud. *Limulus* swims occasionally, lying on its back with the beating gill flaps providing propulsion to a creditable back stroke. The abdomen also has six pairs of dorsal pits, which mark the position of internal muscle attachments.

The tail of *Limulus,* called the telson, is not a true telson because it does not bear the anus. While folklore holds it to be a formidable weapon, the telson actually is used by *Limulus* for digging and for righting itself.

The gills of *Limulus* (see Fig. 257) are home to the triclad flatworm *Bdelloura candida* (see Fig. 148). *Limulus* blood extracts are in clinical use as the basic ingredient in a diagnostic test for endotoxins produced by gram-negative bacteria.

FIG. 253

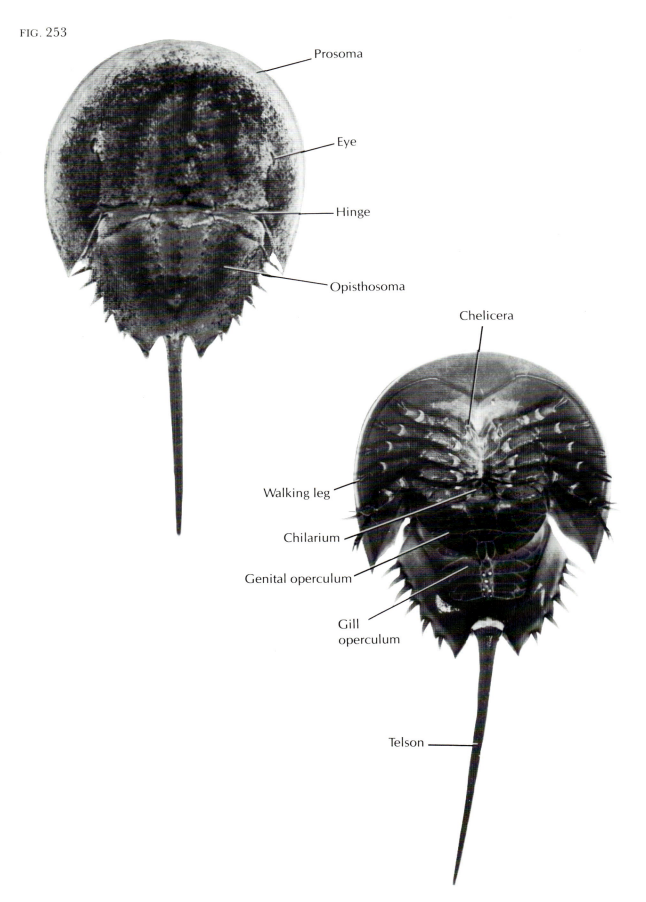

Prosoma

Eye

Hinge

Opisthosoma

Chelicera

Walking leg

Chilarium

Genital operculum

Gill
operculum

Telson

FIG. 254

FIG. 255. Ventral view of the prosoma of a mature male *Limulus*. The chelae on the first pair of walking legs are modified into claspers.

FIG. 256. Close-up of the gnathobases at the distal ends of the walking legs. These large, heavily spined segments masticate food before it enters the mouth.

FIG. 257. Male *Limulus* in which the genital operculum has been reflected anteriorly. The gill lamellae of the following segment are uncovered, as are the male gonopores. The latter structures are round and elevated on papillae. The corresponding female openings are transverse slits that are not elevated.

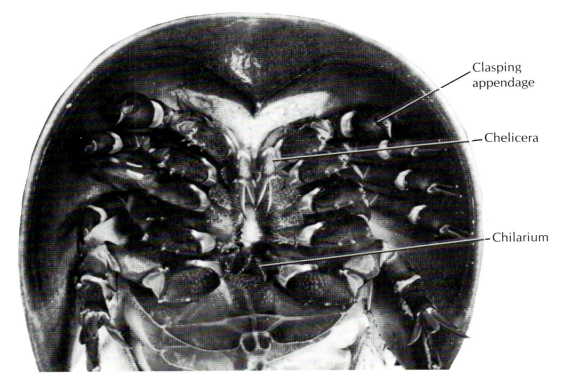

Clasping
appendage

Chelicera

Chilarium

FIG. 255

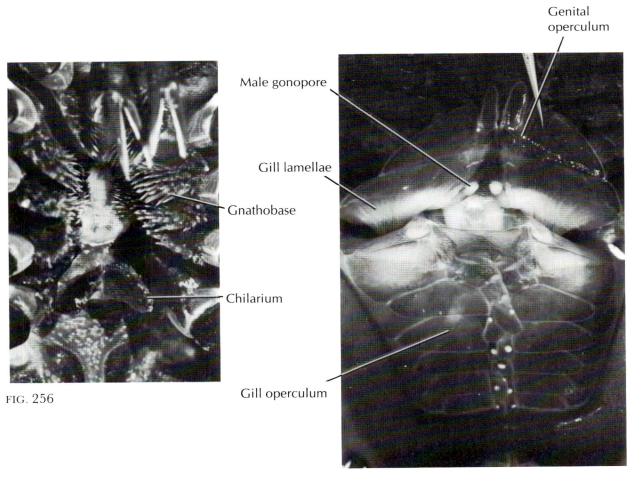

Gnathobase

Chilarium

FIG. 256

Genital
operculum

Male gonopore

Gill lamellae

Gill operculum

FIG. 257

FIG. 259. Specimen shown in Fig. 258 with heart removed to show the underlying digestive system. The esophagus extends anteriorly from the mouth to the proventriculus (crop), which arches dorsally and posteriorly, followed in sequence by the gizzard and stomach-intestine. This last connects to a short rectum in the posterior reaches of the opisthosoma, and the anus is located just anterior to the base of the telson on the ventral side of the animal. The anterior half of the stomach-intestine is supported ventrally by a cartilagelike shelf called the endocranium or, occasionally, the endosterite. This skeletal structure separates the esophagus and the ventral nerve cord from the overlying stomach-intestine.

FIG. 260. The *Limulus* brain, shown in dissection here, consists of a circumoral nerve ring. (In this specimen, the esophagus has been removed.) The nerve ring is composed of several fused ganglia that lie inside the vascular ring surrounding the mouth. The nerve ring gives off several branches, including the ventral nerve cord (see Fig. 261), that are all encased in blood vessels.

Proventriculus   Gizzard   Stomach-intestine

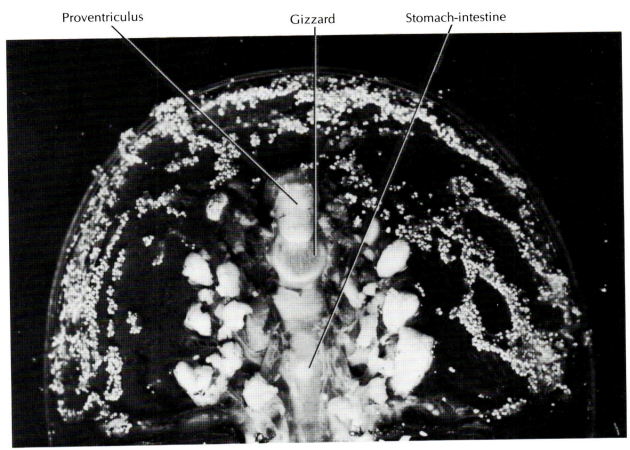

FIG. 259

Endocranium   Vascular ring
containing nerve
ring

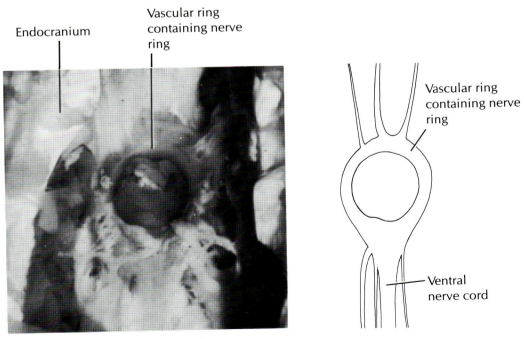

Vascular ring
containing nerve
ring

Ventral
nerve cord

FIG. 260

FIG. 261. Same specimen as in Figs. 258 and 259 with the heart, stomach-intestine, and endocranium removed to reveal the ventral nerve cord. The nerve cord, which is actually a double cord as in all arthropods, runs inside a blood vessel for its entire length. There are ganglia along the cord from which lateral and dorsal nerves branch.

FIG. 262. LM (×17) of the trilobite larva of *Limulus*. Horseshoe crabs often mate in spectacular congregations on intertidal beaches. The female usually digs a hole in the sand for the eggs. Then, the male, which clasps the abdomen of the female with the modified walking legs (see Fig. 255), releases sperm over the eggs as they are shed. The embryos are encased in a leathery capsule from which the trilobite larvae eventually hatch. The telson gradually emerges during subsequent molts.

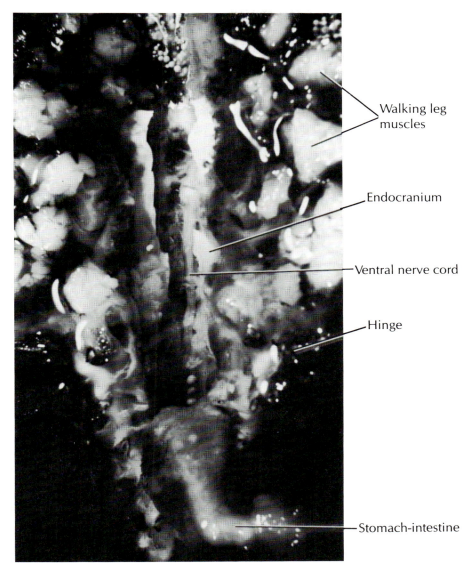

Walking leg
muscles

Endocranium

Ventral nerve cord

Hinge

Stomach-intestine

FIG. 261

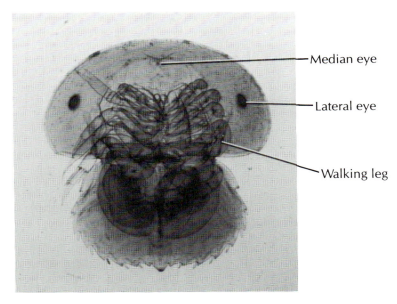

Median eye

Lateral eye

Walking leg

FIG. 262

FIG. 263. The pycnogonids are an arthropod group whose taxonomic position is problematic. The sea spiders, such as the unidentified species shown in this LM, share several anatomical characteristics with the chelicerates: chelicerae, brain, sense organs, and circulatory system. On the other hand, the sea spiders' segmented body, multiple gonopores, additional pairs of walking legs (some sea spiders have ten or twelve pairs of walking legs) are not chelicerate characteristics. Some authors have allied sea spiders with arachnids; others claim there is no relationship. The late P. A. Meglitsch wrote: "If one were asked to design a thing from outer space, one might do worse than take a pycnogonid as his model." And, almost simultaneously, "Pycnogonids are queer animals, with queer habits." They are found only in the marine environment. Many are carnivorous, usually eating coelenterates or bryozoans.

The specimen shown here (×20) is male. It is carrying egg masses on its ovigerous legs. The ovigerous legs are reduced in females, or lacking entirely. Pycnogonid internal anatomy is not complicated. There are no excretory or respiratory systems. The circulatory system is simple and typically arthropod. The heart lies in a blood-filled pericardial sinus, and blood enters the heart through ostia. According to most accounts, digestion in sea spiders is remarkable: some of the cells lining the wall of the intestine detach and, while floating free in the lumen, phagocytize food particles. The cells reattach to the intestine, transfer nutrients to other cells, and then detach again to expel digestive wastes.

FIG. 264. Dorsal view (LM, ×80) of the anterior end of a sea spider.

FIG. 265. Higher-magnification LM (×175) of the ventral side of the head of a sea spider.

FIG. 266. The development of pycnogonids is unremarkable and is typically arthropod. The protonymphon larva is shown here (SEM, ×80). Three pairs of appendages are present when the larva hatches. Additional appendages appear during subsequent molts.

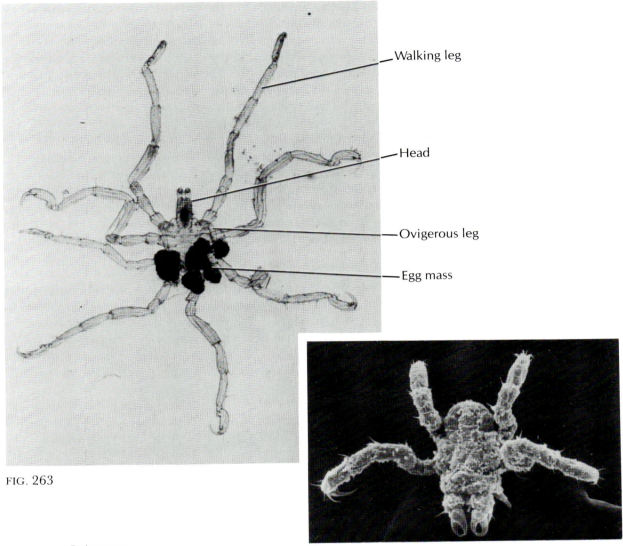

Walking leg

Head

Ovigerous leg

Egg mass

FIG. 263

FIG. 266

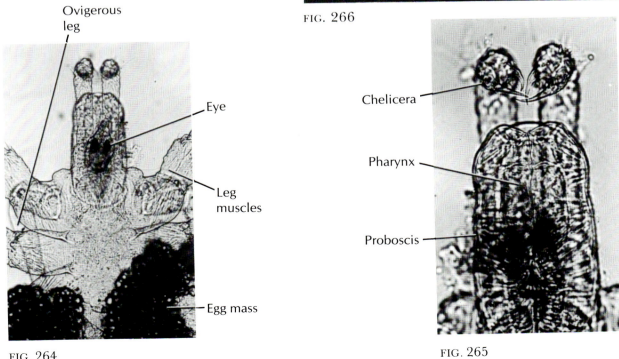

Ovigerous leg

Eye

Leg muscles

Egg mass

FIG. 264

Chelicera

Pharynx

Proboscis

FIG. 265

FIG. 267. Members of the class Branchiopoda are considered to be some of the more primitive crustaceans. The anostracans, or fairy shrimps, such as *Eubranchipus,* shown here (LM, ×20), characteristically have a thorax with eleven to nineteen segments. The thoracic somites bear paired appendages called phyllopods. Usually, the fairy shrimps swim ventral side up, with the phyllopods beating metachronously. The phyllopods also act as gills and food filters in most species. There is no carapace, as the name of the order (Anostraca) implies. The anostracans, like most of the branchiopods, are found only in fresh water.

FIGS. 268 AND 269. *Daphnia,* shown here in LM (×60) (Fig. 268) and SEM (×290) (Fig. 269), is a diplostracan (suborder Cladocera) branchiopod. The water fleas are common in fresh water and a few species are found in the ocean as well. A carapace is present, but it does not enclose the head. The thoracic appendages in the cladocerans usually differ in form from one another, and there are usually only five or six thoracic segments. The well-developed second antennae are used for locomotion. Many cladocerans (in company with other branchiopods) display parthenogenetic development. In some species, males have never been found.

FIG. 270. The ostracods, such as *Cypris,* shown here in LM (×30), are enclosed in a bivalve carapace. The thorax and abdomen are greatly reduced and are not segmented externally. The thorax bears only three pairs of appendages. The head, which is completely inside the carapace, bears four pairs of appendages. The two pairs of antennae are well developed and are used in locomotion. The valves of the carapace are clamlike, although another group of crustaceans, the conchostracans, are known as clam shrimps. A dorsal hinge articulates the two carapace valves, and the valves are closed by a central adductor muscle.

The ostracods are common inhabitants of both marine and freshwater environments. A few species live in the soil in wet forests. Many are filter feeders, a few are parasitic, and a few are predaceous.

FIG. 267

First antenna

Second antenna

Phyllopod

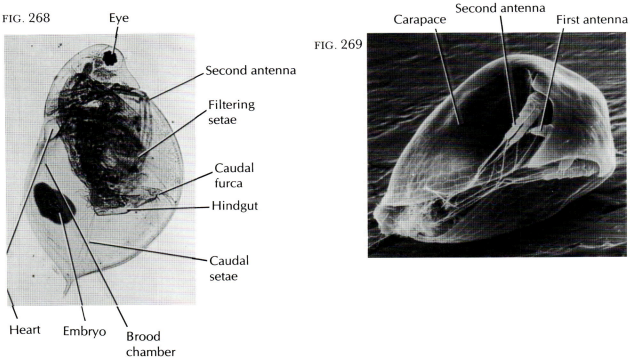

FIG. 268

Eye

Second antenna

Filtering setae

Caudal furca

Hindgut

Caudal setae

Heart

Embryo

Brood chamber

FIG. 269

Carapace

Second antenna

First antenna

FIG. 270

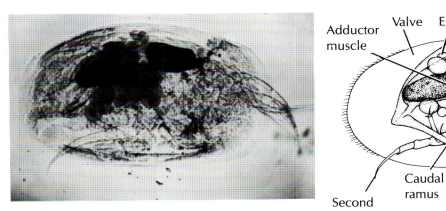

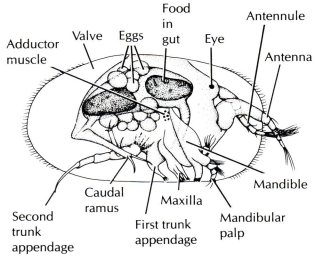

Adductor muscle

Valve

Eggs

Food in gut

Eye

Antennule

Antenna

Second trunk appendage

Caudal ramus

First trunk appendage

Maxilla

Mandibular palp

Mandible

FIGS. 271 AND 272. The Copepoda are a fairly large class of small crustaceans. They are predominately marine, but there are many freshwater species. *Cyclops,* shown in LM (×150) (Fig. 271), and *Mesocyclops edax,* shown in SEM (×90) (Fig. 272), are freshwater forms. Both of these specimens are female. The carapace covers the head and the first one or two thoracic somites; thus, it is neither a head nor a cephalothorax. The head has the usual five pairs of crustacean appendages (antennules, antennae, mandibles, first and second maxillae), with the antennules usually well developed for swimming. A single median eye, derived from the eye of the nauplius larva (see Fig. 297), is usually present. The appendages of the first thoracic segment, called maxillipeds, often function as accessory mouth parts. The appendages of the rest of the thorax are identical thoracic legs, except on the last segment, which is the genital somite (considered by some to be the first abdominal segment). There are no appendages on this segment in the female copepods. During mating, a sac forms on the somite that contains spermatophores in males and fertilized eggs in females (Fig. 272). The abdominal segments lack appendages. The anus is located on the telson and a pair of caudal rami forms the furca.

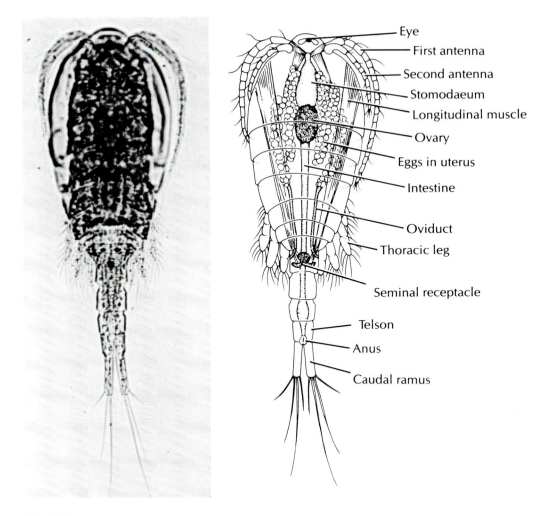

- Eye
- First antenna
- Second antenna
- Stomodaeum
- Longitudinal muscle
- Ovary
- Eggs in uterus
- Intestine
- Oviduct
- Thoracic leg
- Seminal receptacle
- Telson
- Anus
- Caudal ramus

FIG. 271

FIG. 272

FIGS. 273 AND 274. There are two types of barnacles, unstalked and stalked; *Lepas anatifera,* shown here, is stalked. The skeletal plates are calcareous, which is why barnacles were originally classified with the molluscs. The body of barnacles is very much modified from the usual crustacean plan. Barnacles attach to the substratum by their large, prominent head, which is glued in place with a cement secreted from a gland that opens near the antennules. The antennae are entirely absent. The thorax has six pairs of legs that are thrust from openings in the shell plates (Fig. 273). Each leg ends in two thin cirri (hence the name of the class, Cirripedia) that wave through the water to filter out food particles. There is no abdomen. The carapace is a fleshy mantle that lines the inner surface of the shell and forms a mantle cavity around the body (Fig. 274).

Barnacles are exclusively marine. They attach to a wide array of substrates and are one of the more important groups of fouling organisms. Some barnacles are parasitic, of which one of the better-known is *Sacculina,* which parasitizes blue crabs. The adults of the parasitic species often lose all resemblance to free-living barnacles.

FIG. 273

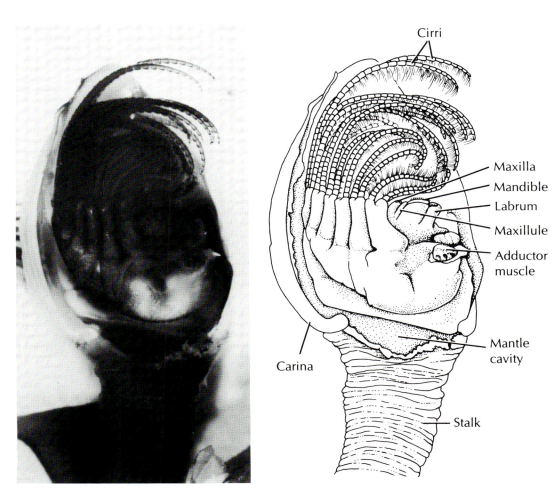

FIG. 274

FIGS. 275–278. The isopods, one of the three main groups of class Malacostraca, are dorsoventrally flattened, have no carapace, and have uniramous walking legs (Figs. 276 and 278) and abdominal gills. In many species, there is a tendency for the abdominal segments to be fused. Isopods are widely distributed in the marine environment. The unidentified species shown in Figs. 275 and 276 are marine forms. Freshwater species exist, and the terrestrial isopods, such as *Oniscus asellus,* shown here in dorsal (Fig. 277) and ventral (Fig. 278) views, are the largest group of crustaceans that live on land.

FIG. 275

FIG. 276

Abdomen

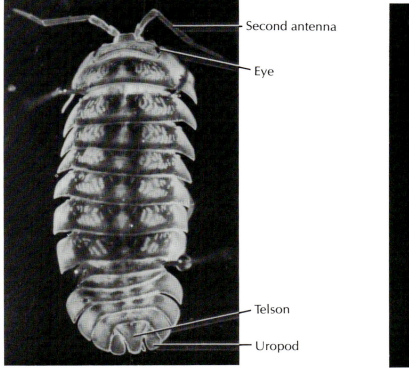

FIG. 277

Second antenna

Eye

Telson

Uropod

FIG. 278

FIG. 279. The amphipods are a second major group of class Malacostraca. They are mostly marine, but there are many freshwater species and a few terrestrial species. Amphipods share several anatomical features with the isopods. They have no carapace, the eyes are unstalked, the thoracic legs are uniramous, and some of the abdominal segments are fused. Unlike the isopods, the amphipods are laterally flattened and the gills are thoracic. Typical amphipod anatomy is seen in this LM (×15) of *Gammarus*.

FIG. 280. Many amphipods live in tubes. *Jassa,* shown here, is an example. It builds open-ended mud tubes on a variety of substrates. In some areas, *Jassa* is an important fouling organism. The second gnathopod in *Jassa* is highly developed. Usually one or both pairs of gnathopods (the second and third thoracic appendages) in the amphipods are subchelate.

FIG. 281. The abdomen is vestigial in the caprellid amphipods, also called skeleton shrimp, such as *Caprella geometrica* shown here (SEM, ×40). The body of these amphipods is modified for climbing. The raptorial appendages allow the animal to cling very tightly to hydroid or bryozoan colonies or algae. *Caprella* attaches to the substrate with the posterior thoracic claws, extends itself from the substrate, and catches passing prey with the gnathopods.

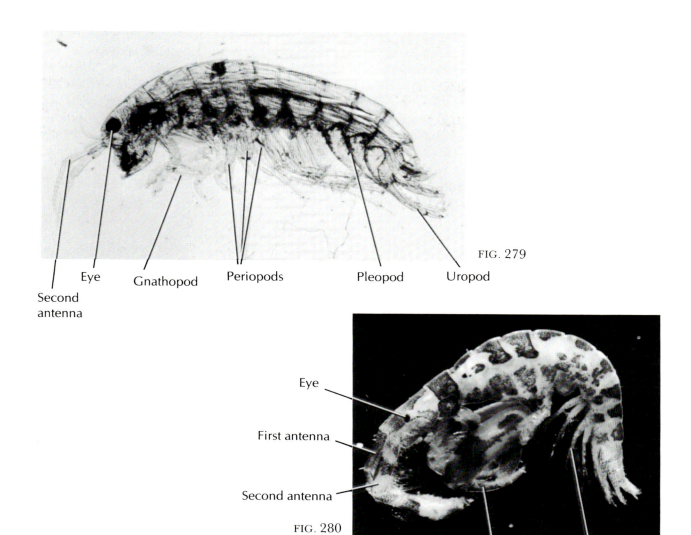

Eye

Gnathopod

Periopods

Pleopod

Uropod

FIG. 279

Second
antenna

Eye

First antenna

Second antenna

FIG. 280

Gnathopod

Pleopod

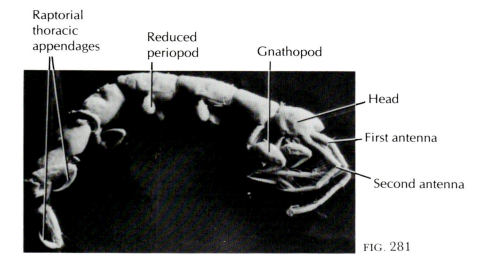

Raptorial
thoracic
appendages

Reduced
periopod

Gnathopod

Head

First antenna

Second antenna

FIG. 281

FIGS. 282–285. By far the largest of the crustacean orders is Decapoda, which includes shrimps, crabs, lobsters and the like. All are well known to both biologists and laymen not only for their obviousness in the environment, but also as gastronomic delights. The crayfish shown here (Fig. 282) is typical of the basic body plan of the many species of crayfish belonging to several genera. The cephalothorax (Fig. 283) is covered by a hard carapace with an anterior rostrum. The carapace forms a lateral gill cover as well, called the branchiostegite. There are six abdominal segments and a telson. The appendages are typically crustacean, except that the first pair of periopods (thoracic walking legs) are developed into large pincers called chelipeds. The first two abdominal appendages (pleopods) are often called gonopods, and the pleopods are called swimmerets (Fig. 284). The last abdominal segment bears uropods (Fig. 285).

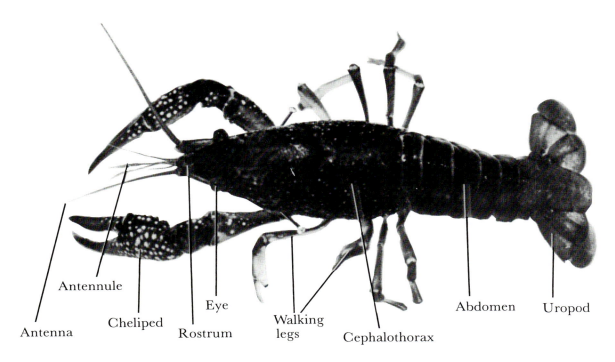

Antennule

Antenna

Cheliped

Rostrum

Eye

Walking legs

Cephalothorax

Abdomen

Uropod

FIG. 282

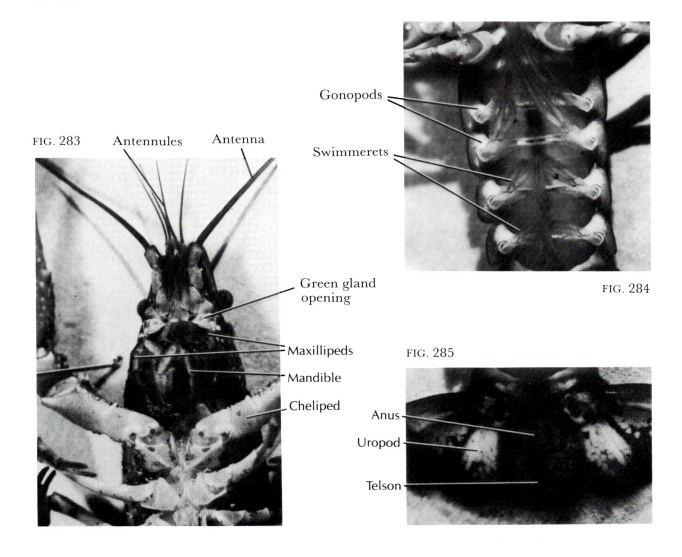

FIG. 283   Antennules   Antenna

Green gland opening

Maxillipeds

Mandible

Cheliped

Gonopods

Swimmerets

FIG. 284

FIG. 285

Anus

Uropod

Telson

FIGS. 286–289. Several views of a dissected crayfish (Figs. 286–288) and a diagram of the anatomy (Fig. 289). Immediately under the dorsal exoskeleton are the heart, gills, and, in this specimen, developing eggs (Fig. 286). The heart and gills have been cut away in Fig. 287 to reveal the underlying organs, and the dorsal exoskeleton of the abdomen has been removed to reveal the abdominal musculature and the intestine. In Fig. 288, the intestine and most of the musculature have been removed from the abdomen to show the ventral nerve cord. The cord is a paired structure, as are the ganglia. The connectives that lead off from the ganglia are clearly seen. (Dissection by L. M. Rowland, University of Maryland.)

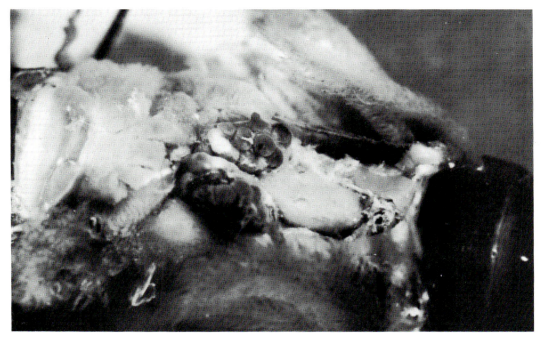

FIG. 286

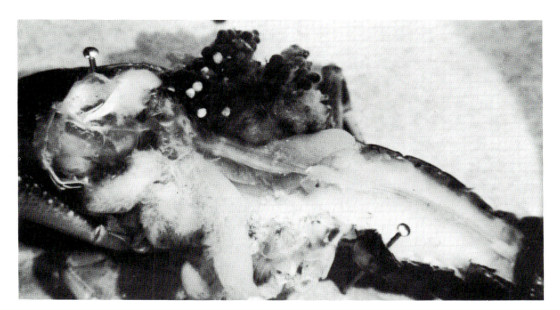

FIG. 287

FIG. 288

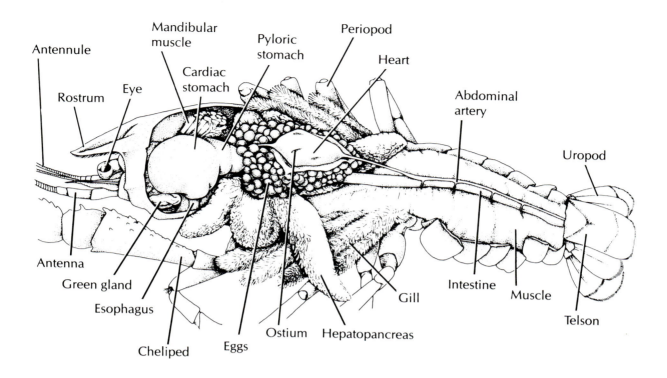

Antennule

Mandibular muscle

Rostrum

Cardiac stomach

Eye

Pyloric stomach

Periopod

Heart

Abdominal artery

Uropod

Antenna

Green gland

Esophagus

Cheliped

Eggs

Ostium

Hepatopancreas

Gill

Intestine

Muscle

Telson

FIG. 289

FIGS. 290–292. The brachyuran crabs, such as *Cancer irroratus,* shown here, have shortened bodies. The cephalothorax is covered by a broad, flat carapace (Fig. 290). The antennae are usually short, and the third maxillipeds are flat, forming a covering over the mouth and the other mouth parts (Fig. 292). The periopods are well developed; the first pair consists of chelipeds, while the third pair lacks chelae (Figs. 290 and 291). The abdomen is reduced and folded under the cephalothorax. In males, the abdomen is very narrow, and of the abdominal appendages, only the copulatory pleopods remain. The abdomen is broader in the female, and four pairs of pleopods, which are often highly branched (see Fig. 294), are used to brood the eggs.

There is a great deal of anatomical diversity among the brachyurans, but an understanding of the typical anatomical plan will aid in interpreting the anatomy of any crab.

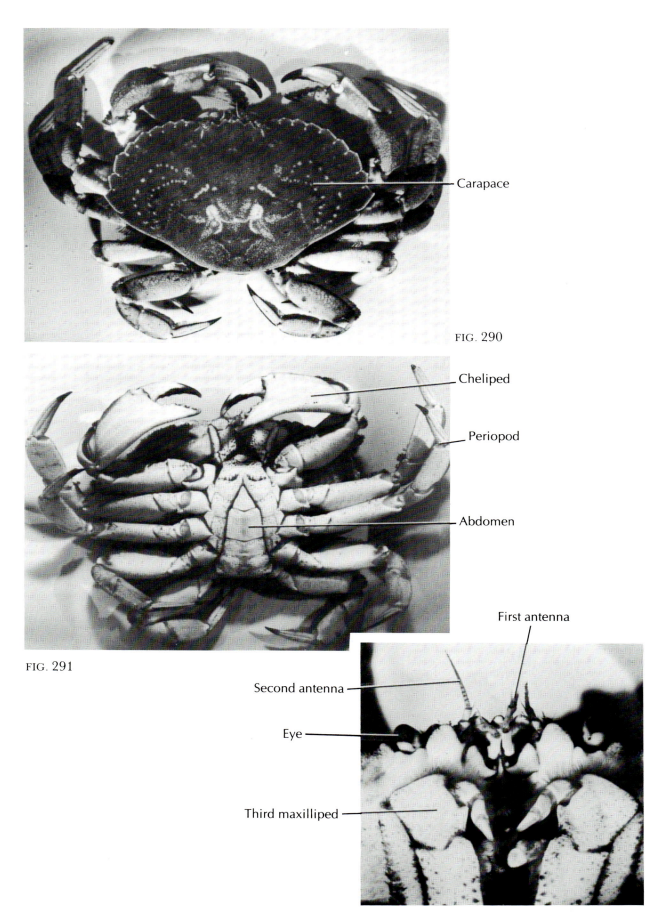

Carapace

FIG. 290

Cheliped

Periopod

Abdomen

FIG. 291

First antenna

Second antenna

Eye

Third maxilliped

FIG. 292

FIG. 293. Most crabs do not swim; however, crabs in the family Portunidae are excellent swimmers. The last pair of periopods is modified into a swimming leg, terminating in a broad paddle, shown here on the blue crab *Callinectes sapidus*.

FIG. 294. Close-up of the pleopods of a female *Callinectes*. The abdomen has been reflected back from the underside of the cephalothorax.

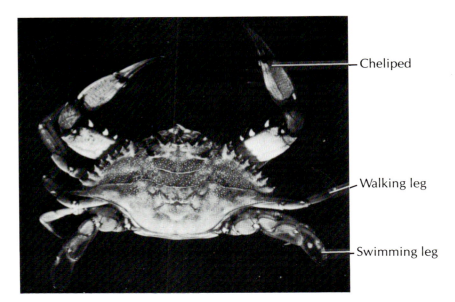

Cheliped

Walking leg

Swimming leg

FIG. 293

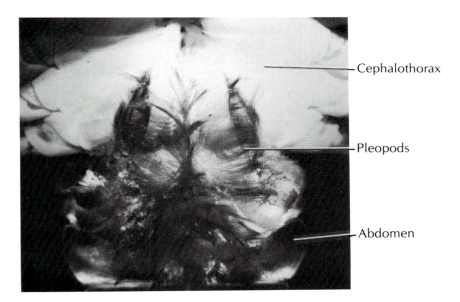

Cephalothorax

Pleopods

Abdomen

FIG. 294

FIG. 295. Dorsal dissection of a male *Cancer borealis*. The carapace and hypodermis have been removed in Fig. 295a to reveal the underlying organs. The heart, stomach, and digestive gland have been removed in Fig. 295b to show the gonads. The anatomy is diagrammed in Fig. 295c.

FIG. 295a

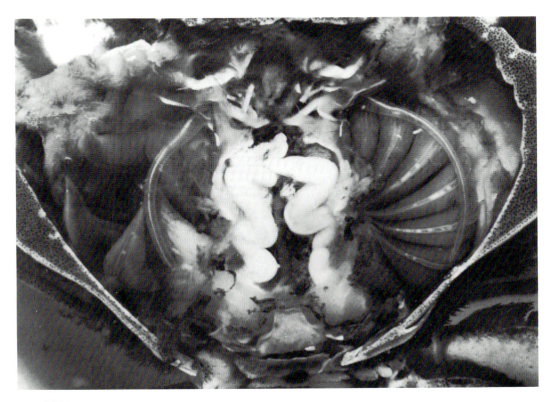

FIG. 295b

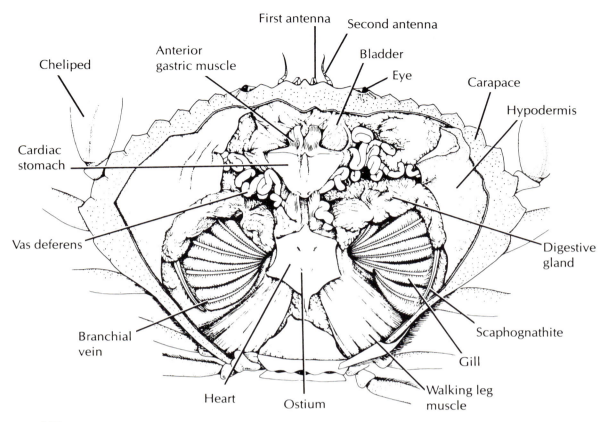

Cheliped

Anterior
gastric muscle

First antenna   Second antenna

Bladder

Eye

Carapace

Hypodermis

Cardiac
stomach

Vas deferens

Digestive
gland

Branchial
vein

Scaphognathite

Gill

Heart   Ostium   Walking leg
muscle

FIG. 295c

FIG. 296. The brachyurans are the best known of the crabs. However, some other decapods are commonly encountered that show interesting anatomical deviations from the typical crustacean body plan. For example, the body of *Emerita talpoida,* the mole crab, shown here in dorsal (Fig. 296a) and ventral (Fig. 296b) views, is highly modified for burrowing. The abdomen is tucked under the cylindrical body, chelipeds are absent, and the thoracic appendages and uropods are flattened into very effective digging tools. The eyes, on long stalks, protrude from the sand along with the first and second antennae.

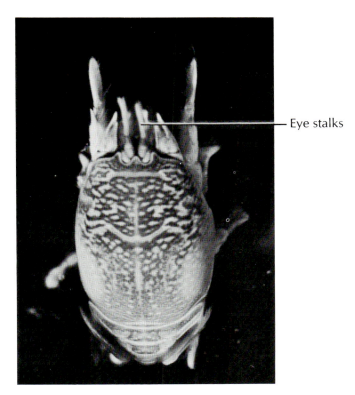

————Eye stalks

FIG. 296a

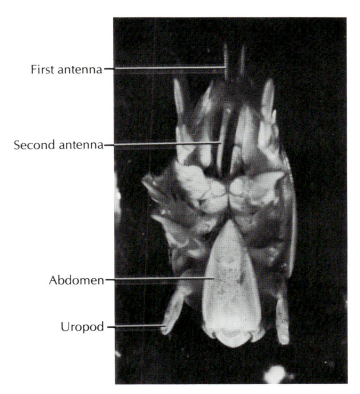

First antenna————

Second antenna————

Abdomen————

Uropod————

FIG. 296b

FIGS. 297–300. The larval history of most crustacean species is complex. There are usually several stages and each stage has a strikingly different morphology. The anatomical changes occur during the molts. The basic larval type is the nauplius (Fig. 297, ×85), which is present in a large number of crustacean taxa. In the Malacostraca, the protozoea and then the zoea (Fig. 298, ×18) are the stages following the nauplius. The zoea is easily recognized in most species by the long rostral and lateral spines on the carapace. In the decapoda, the zoea is often followed by the shrimplike mysis (Fig. 299, ×14), which transforms into the adult in most of the shrimp and lobster groups. In many of the crabs, the mysis transforms into a megalopa larva (Fig. 300, ×16), which finally molts into the adult form. The diversity in crustacean larval forms supports the idea that the larval stages reflect evolutionary relationships, perhaps an excellent example of ontogeny actually recapitulating phylogeny.

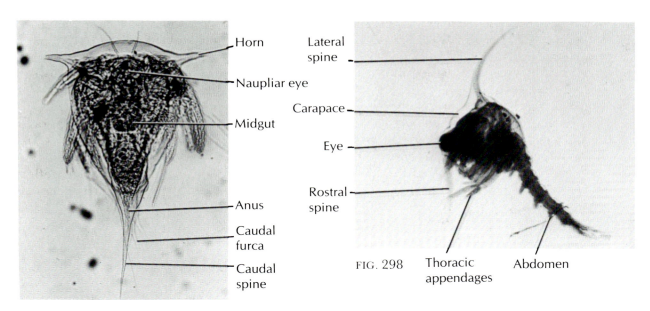

Horn

Naupliar eye

Midgut

Anus

Caudal furca

Caudal spine

FIG. 297

Lateral spine

Carapace

Eye

Rostral spine

FIG. 298

Thoracic appendages

Abdomen

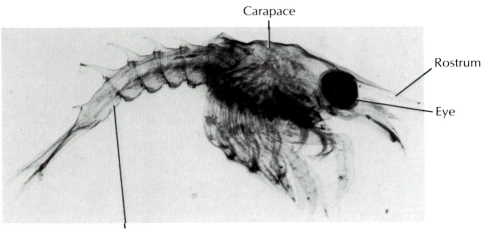

Carapace

Rostrum

Eye

FIG. 299

Abdomen

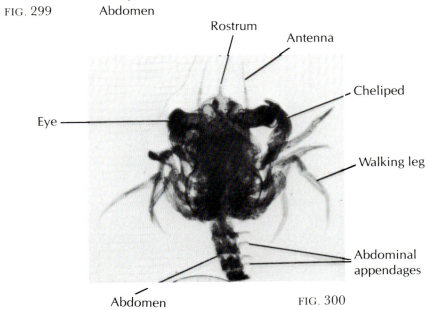

Rostrum

Antenna

Cheliped

Eye

Walking leg

Abdominal appendages

Abdomen

FIG. 300

SECTION X
Taxonomic Summary

Phylum Echinodermata
CLASS STELLEROIDEA
  SUBCLASS ASTEROIDEA
   Order Forcipulatida
      Asterias forbesi
      Pisaster ochraceus
  SUBCLASS OPHIUROIDEA
   Order Ophiurida
      Ophioderma brevispina
CLASS ECHINOIDEA
  SUBCLASS EUECHINOIDEA
   SUPERORDER ECHINACEA
   Order Arbacioida
      Arbacia punctulata
   Order Temnopleuroida
      Lytechinus pictus
   Order Echinoida
      Strongylocentrotus droebachiensis
   SUPERORDER GNATHOSTOMATA
   Order Clypeasteroida
      Clypeaster subdepressus
      Mellita quinquiespreforata
      Dendraster excentricus
   SUPERORDER ATELOSTOMATA
   Order Spatangoida
      Moira atropos
CLASS HOLOTHUROIDEA
   Order Dendrochirotida
      Thyone briareus

FIG. 301. The echinoderms are the only major deuterostome invertebrate phylum. The sea star, *Asterias forbesi,* shown in dissection here, displays many of the typical characteristics of the echinoderms. Although the larvae of the various echinoderm groups are bilaterally symmetrical (see Figs. 323–327), the adults display pentameric radial symmetry. The radial symmetry is usually viewed as a secondarily derived characteristic rather than a persistent primitive trait. Echinoderms have a calcareous internal skeleton composed of articulating plates (sea stars and brittle stars), a solid fused test (urchins), or tiny ossicles (sea cucumbers). Parts of the skeleton often protrude from the body surface as spines. Locomotion in the echinoderms is by means of tube feet (podia) (see Fig. 305) that are powered by the water vascular system. This water vascular system, an extension of the coelom, is often called the ambulacral system.

The nervous system in asteroids is fairly diffuse and mostly associated with the epidermis. A circumoral nerve ring is present but lacks ganglia. The only sense organs are eyespots at the tips of the tentacles.

The coelomic fluid serves the circulatory function. The peritoneum is ciliated and the beating cilia continually circulate the coelomic fluid. Gas exchange and nitrogen excretion occur across the tube feet and also across very thin projections of the aboral body wall, the papulae or dermal branchiae (see Fig. 302). Phagocytic coelomocytes congregate in the tips of some papulae, which then pinch off and eject the coelomocytes.

Carnivorous sea stars with flexible arms, such as *Asterias,* feed on bivalve molluscs. The star captures the clam with its tube feet and positions its mouth over the gape of the clam's shell. The sea star then everts its stomach through the mouth and inserts it through the gape in the shell, which has been pulled slightly ajar by the arms and tube feet. Digestion starts inside the bivalve. In sea stars with inflexible arms or with tube feet lacking suckers, the stomach is not usually eversible.

Echinoderms are found exclusively in marine habitats, probably because they lack any sort of excretory system. Asteroids and holothuroids have excellent powers of regeneration and have served as models in the study of the processes of wound healing.

FIG. 301

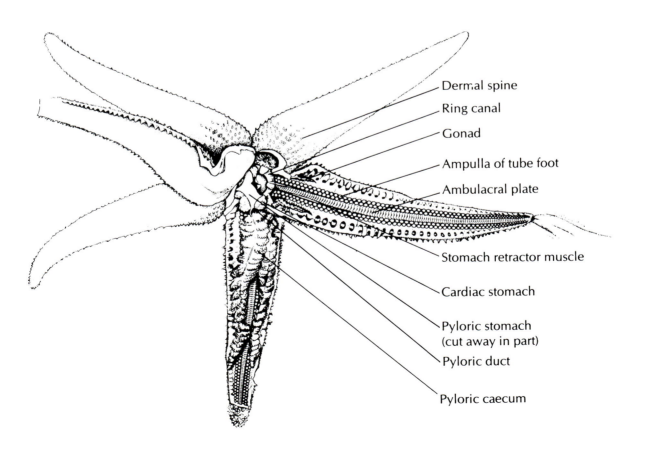

Dermal spine

Ring canal

Gonad

Ampulla of tube foot

Ambulacral plate

Stomach retractor muscle

Cardiac stomach

Pyloric stomach
(cut away in part)

Pyloric duct

Pyloric caecum

FIG. 305

FIG. 306. SEM (×110) of the skeleton of a dermal spine of *Asterias*. The smaller pedicellariae that surround the spine are visible in the foreground. The epithelium has been removed by sonication.

FIGS. 307 AND 308. The pedicellariae of *Asterias,* shown here in SEMs, are on contractile stalks. The two jaw ossicles are articulated with one another and muscles open and close the jaws like scissors. The function of the pedicellariae seems to be to keep the surface of the sea star clean. The covering epithelium in Fig. 307 (×270) of the pedicellaria is intact. The epithelium has been removed from the pedicellaria in Fig. 308 (×310) to reveal the skeleton and musculature.

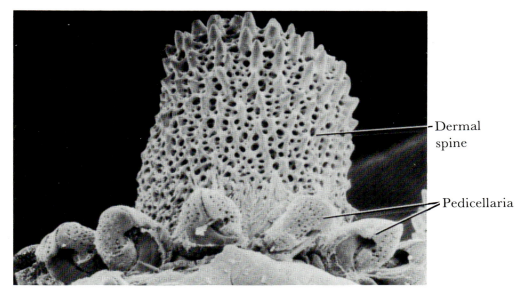

FIG. 306

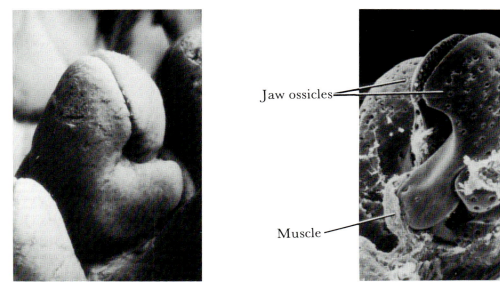

FIG. 307

FIG. 308

FIG. 309. The ophiuroid echinoderms, the brittle stars, superficially resemble the asteroids. However, there are several important differences. The arms and the central disc are much more obvious in the brittle stars, as in the anesthetized *Ophioderma brevispina* shown here. The arms are highly mobile; they hold the disc above the substrate and move the animal rapidly over the sea floor with a rowing motion.

FIG. 310. The disc of brittle stars is covered with plates. The oral surface of *Ophioderma*, shown here, has several large plates (called shields) that surround the mouth. These shields, with their toothlike appendages, serve as jaws. One of the oral shields is the madreporite, which in the brittle stars is pierced by a single pore. Pedicellariae and papulae are absent.

FIG. 311. The podia, reduced in ophiuroids, are not used for locomotion. The ambulacral groove, so obvious in asteroids, does not occur on the oral side of the arms, as shown in this close-up of *Ophioderma*. Instead, small tube feet emerge between the lateral and oral plates on the arms. The podia are thought to be primarily sensory in the brittle stars, although they do pass adherent food particles toward the mouth.

FIG. 312. The skeletal plates on the *Ophioderma* arms interdigitate like vertebrae, as seen in this SEM ($\times$40). There are four sets of plates, one on the oral surface, two on the lateral surface, and one on the aboral surface.

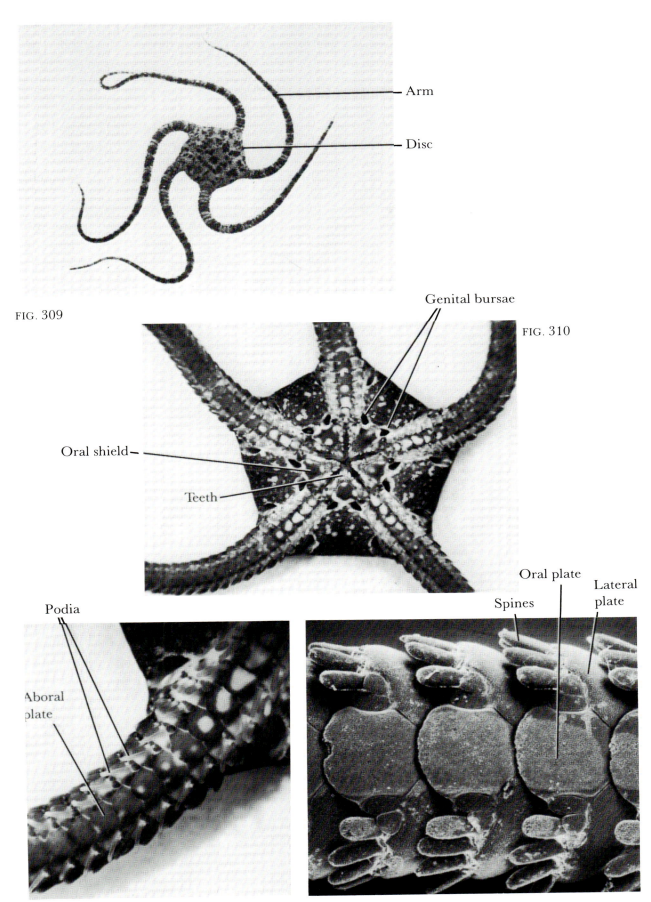

Arm

Disc

FIG. 309

Genital bursae

FIG. 310

Oral shield

Teeth

Oral plate

Spines

Lateral plate

Podia

Aboral plate

FIG. 311

FIG. 312

FIG. 313. The regular echinoids, such as *Arbacia punctulata,* shown here, are circular or oval in shape. The skeletal plates are fused into a solid structure, the test (see Fig. 315), which is armed with movable spines (Fig. 313a). External gills are often present (Fig. 313b and d). Tube feet are used for locomotion (along with the spines) and assist in feeding. The tube feet are located along five radial sections of the test, called ambulacral sections. The water vascular system is very similar to that of the asteroids.

Most sea urchins have two or more types of spines, but *Arbacia* has only one type. Pedicellariae are scattered over the external surface of the test. Some sea urchin species have pedicellariae equipped with poison sacs. The pedicellariae are stalked as in the asteroids, but they are usually three-jawed, as seen in this SEM (×840) (Fig. 313c) of *Lytechinus pictus. (Photograph by D. B. Bonar, University of Maryland.)*

Arbacia and some other regular urchins are frequent subjects of embryological studies. They can be induced to spawn with an injection of potassium chloride. The sperm released rapidly fertilize the eggs, which soon begin to develop. (Fig. 313d redrawn from F. A. Brown, Jr., *Selected Invertebrate Types* [New York: Wiley, 1949], Fig. 217 on p. 529.)

FIG. 313c

FIG. 313a

FIG. 313b

FIG. 314. The internal anatomy of sea urchins is relatively simple, as seen here in *Arbacia punctulata*. The test has been cut in half horizontally between the oral and aboral poles, so the perspective is down into the oral region. The gut forms a frilly balcony attached to the inside of the test (Fig. 314a). It has been removed in Fig. 314b to display the underlying structures, which include the ampullae of the podia and the complex musculature and skeleton of Aristotle's lantern, the feeding organ. The diagram (Fig. 314c) shows a cutaway section of the gut that exposes the underlying structures.

FIG. 314a

FIG. 314b

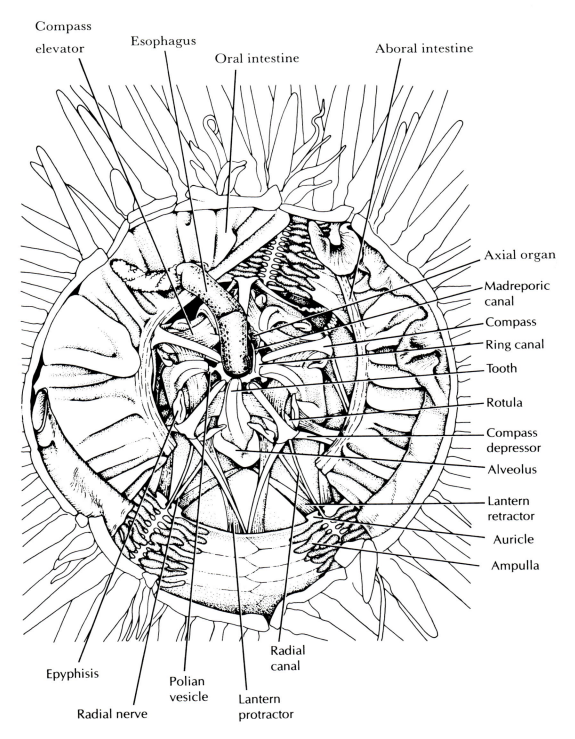

Compass
elevator

Esophagus

Oral intestine

Aboral intestine

Axial organ

Madreporic
canal

Compass

Ring canal

Tooth

Rotula

Compass
depressor

Alveolus

Lantern
retractor

Auricle

Ampulla

Epyphisis

Radial nerve

Polian
vesicle

Lantern
protractor

Radial
canal

FIG. 314c

FIG. 315. Pentameric radial symmetry is clearly evident in the tests of the regular urchins. The aboral surface of a cleaned test of *Strongylocentrotus droebachiensis* is shown here. The tube feet protrude through the holes in the ambulacral plates. The tubercles on the test articulate with the base of the spines.

FIGS. 316–318. The clypeasteroid (sand dollars, *Mellita quinquiespreforata*, Fig. 316; sea biscuits, *Clypeaster subdepressus*, Figs. 317 and 318) is one of the irregular urchin types. Clypeasteroids are usually round or oval but greatly flattened in the oral–aboral axis. The spines are short and fine compared with those of regular urchins. The ambulacral areas are reduced to the five-rayed petaloids on the aboral surface. The podia in the petaloids are not used for locomotion; their function seems to be entirely respiratory. Small podia are located on the aboral surface of the clypeasteroids and assist the spines in locomotion. The anus and periproct are on the oral surface and their location imparts a secondary bilateral symmetry to the organism (Fig. 318). Several of the sand dollar genera, including *Mellita* (Fig. 316), have large holes through the test called lunules, which also impart bilateral symmetry. The peristome is at the center of the oral surface in the clypeasteroids (Fig. 318). The madreporite and the genital pores are on the aboral surface at the center of the petaloids.

FIG. 319. The other irregular urchin type is the spatangoid, such as the heart urchin *Moira atropos*, shown in both oral and aboral perspective here. The spatangoids have the greatest degree of bilateral symmetry in the Echinoidea. The peristome, which is supported by a number of small plates, is well forward on the aboral surface and is bordered by a liplike projection, the labrum. The oral ends of the ambulacra, called phyllodes and containing podial pores, expand around the peristome. The two long ambulacra that extend posteriorly from the phyllodes enclose a flat interambulacrum, the plastron. All of these features are evident on the *Moira* test on the left in Fig. 319, which was photographed with the aboral surface up. The plastron extends posteriorly to the periproct (also supported by plates, which are missing from this specimen). The arched aboral surface contains four, fairly short, narrow petaloids and one deeply grooved anterior ambulacrum that runs toward the peristome. The spatangoids move about using the short, fine spines that cover the test. Locomotory podia are absent. Branchial podia are present in the petaloids, and buccal podia occur on the phyllodes.

FIG. 315

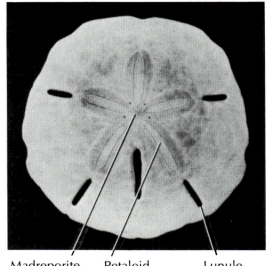

FIG. 316

Madreporite   Petaloid   Lunule

Gonopore   Petaloid

FIG. 317

Peristome

FIG. 318

Periproct

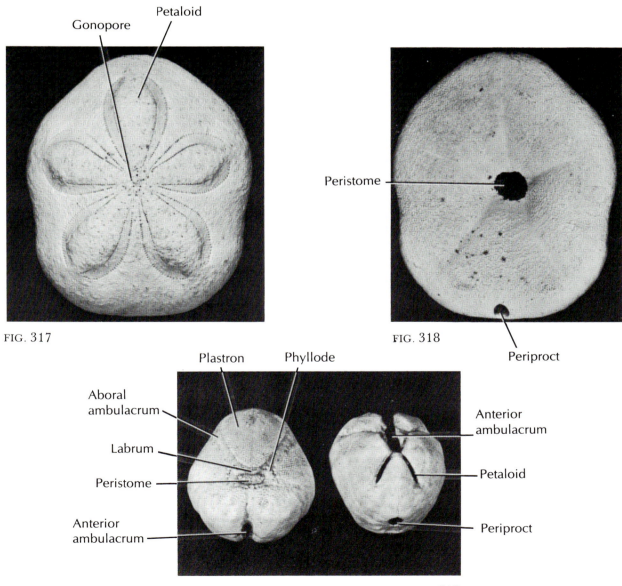

Plastron   Phyllode

Aboral
ambulacrum

Labrum

Peristome

Anterior
ambulacrum

Anterior
ambulacrum

Petaloid

Periproct

FIG. 319

FIG. 320. The holothurian sea cucumbers, such as *Thyone briareus,* shown in dissection here, are relatively soft-bodied, unlike the rest of the echinoderms. Skeletal ossicles are present in the holothurians, but they are tiny and not fused. Many of the bottom-dwelling sea cucumbers lie with one side of the body on the substrate, so that three of the ambulacral areas have become concentrated on this so-called ventral surface, or the sole. The sole imparts a secondary bilateral symmetry to the pentameric radial symmetry of the sea cucumber. *Thyone* does not follow this pattern, however; in it, tube feet cover the entire body uniformly. The burrowing sea cucumbers, many of which lack podia entirely, usually display only radial symmetry. Tentacles always encircle the mouth of the holothurians. These structures are modified buccal podia and are connected into the water vascular system. Sea cucumbers have an internal madreporite, which hangs free in the coelom, suspended from the stone canal. Thus, unlike the other echinoderms, coelomic fluid, rather than sea water, courses through the water vascular system.

FIG. 320

FIG. 320. (cont.). Many species of sea cucumber have the peculiar habit of evisceration when they are disturbed. Some species extrude specialized tubules (tubules of Cuvier) through the anus. These structures are attached to the bases of the respiratory trees and are usually covered with an adhesive substance. (The tubules of some species are toxic.) The tubules break off the respiratory trees after extrusion and are eventually regenerated. A more spectacular evisceration occurs in other species. In *Thyone,* the anterior end of the animal blows out. Tentacles, pharynx, gut, and associated structures are all lost and the retractor muscles are severed. Other species eviscerate from the cloacal end, expelling respiratory trees, gut, and gonads. In all these species, the missing organs are regenerated. The expulsion of the Cuverian tubules is clearly a defensive maneuver. The function of the more radical eviscerations is less clear; they occur much more frequently in the laboratory than in the field. Evisceration in *Thyone* can be induced by injecting potassium chloride into the coelomic fluid.

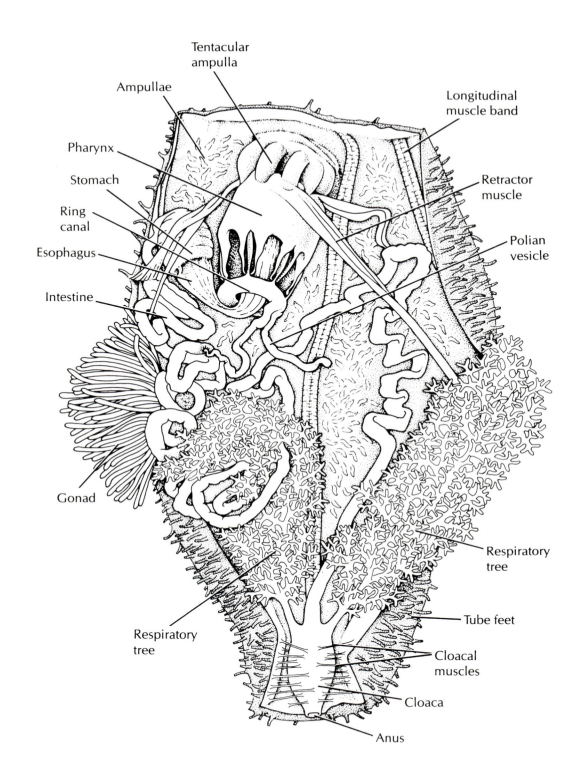

Tentacular
ampulla

Ampullae

Pharynx

Stomach

Ring
canal

Esophagus

Intestine

Gonad

Longitudinal
muscle band

Retractor
muscle

Polian
vesicle

Respiratory
tree

Tube feet

Cloacal
muscles

Cloaca

Anus

Respiratory
tree

FIG. 320 *(cont.)*

FIG. 321. Close-up of the ampullae in a freshly dissected *Thyone*.

FIG. 322. Respiratory trees of *Thyone*.

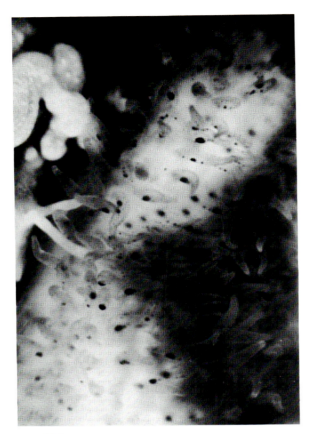

FIG. 321

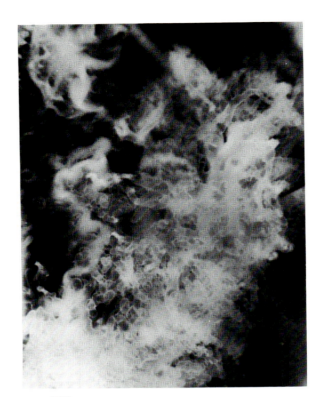

FIG. 322

FIGS. 323–327. The echinoderm larval stages are bilaterally symmetrical and swim in the plankton using ciliated bands surrounding their bodies. There is little resemblance to the adult in echinoderm larval stages.

The SEM (Fig. 324, ×150) shows a young bipinnaria of *Pisaster ochraceus. (Photograph by R. Burke, University of Victoria.)* The asteroid bipinnaria develops into a brachiolaria larva by rearrangement of ciliated bands and arm growth. Brachiolar arms and an adhesive sucker appear (Fig. 325, ×21). This last stage usually attaches to the substrate by the sucker. The posterior portion of the larva, which contains the stomach and stomatocoel portion of the coelom, develops into a tiny starfish; the rest of the larva degenerates.

In the echinoids, the gastrula forms a larva called the pluteus, shown here for *Arbacia* (Fig. 326, LM, ×140) and for the sand dollar *Dendraster excentricus* (Fig. 327, SEM, ×115; *photograph by R. Burke, University of Victoria),* which eventually metamorphoses into a young echinoid.

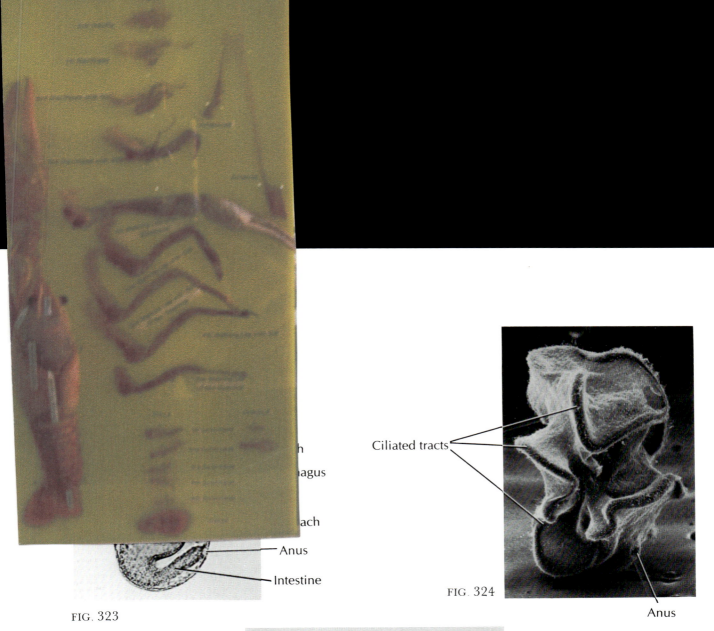

ch

agus

ach

—Anus

—Intestine

FIG. 323

Ciliated tracts

FIG. 324

Anus

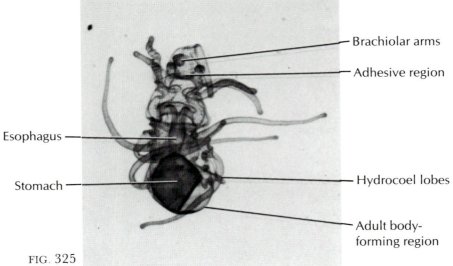

—Brachiolar arms

—Adhesive region

Esophagus —

—Hydrocoel lobes

Stomach —

—Adult body-
forming region

FIG. 325

FIG. 326

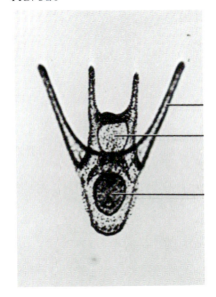

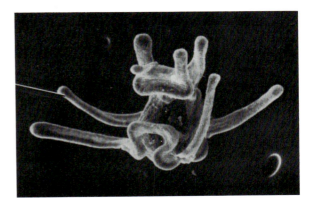

FIG. 327

SECTION XI
Taxonomic Summary

Phylum Chordata
Subphylum Urochordata
CLASS ACIDIACEA
Order Phlebobranchiata
Ciona intestinalis
ORDER STOLIDOBRANCHIATA
Molgula manhattensis
Botryllus schlosseri

FIG. 328. The tunicates, or sea squirts, are classified as chordates even though they lack a backbone. The larval stage (see Fig. 330) possesses a notochord, a dorsal hollow nerve cord, and pharyngeal gill slits. All of these characteristics are lacking in the adults, such as *Ciona intestinalis,* shown here surrounded by a cluster of another tunicate species, *Molgula manhattensis.* These two species exemplify the extremes in tunicate body shape, which ranges from spherical *(Molgula)* to cylindrical *(Ciona).* One end of the body contains two siphons, the atrial and the buccal. The sea squirt is attached to the substrate at the end opposite the siphons. The body is covered by a secreted tunic.

The internal anatomy is fairly simple. The buccal siphon, often with a ring of internal tentacles (see Fig. 329), leads into the pharyngeal basket, a filtering structure lined with ciliary tracts that direct food toward the stomach (the main tract is called the endostyle) and perforated by slits through which water passes out of the basket. The slits are lined with cilia, the action of which moves water through the animal. Water passes from the pharyngeal basket into a chamber called the atrium, then passes over the anus and the openings of the reproductive system, and finally exits through the atrial siphon.

FIG. 329. Close-up of the siphons of *Ciona intestinalis.*

FIG. 330. Tadpole larva of the sea squirt *Botryllus schlosseri* (LM, ×60).

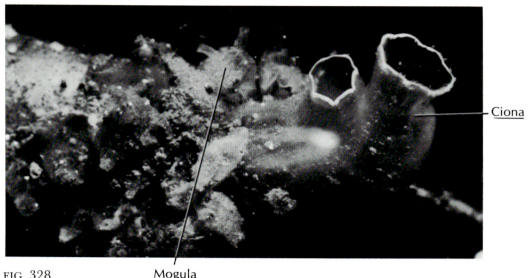

FIG. 328

Ciona

Mogula

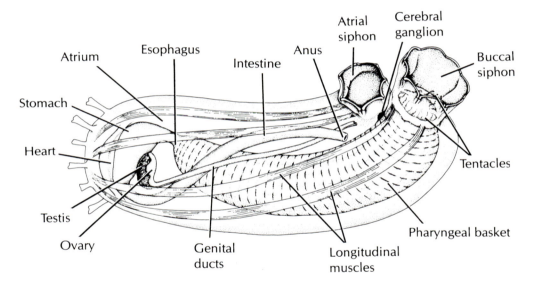

Atrium
Esophagus
Intestine
Anus
Atrial siphon
Cerebral ganglion
Buccal siphon
Stomach
Heart
Testis
Ovary
Genital ducts
Longitudinal muscles
Tentacles
Pharyngeal basket

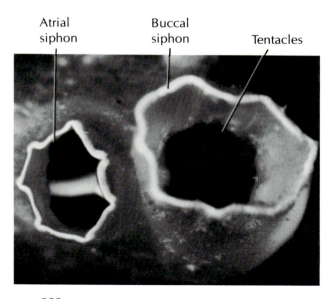

Atrial siphon
Buccal siphon
Tentacles

FIG. 329

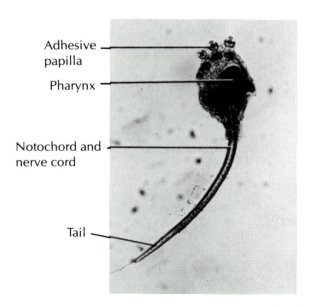

Adhesive papilla
Pharynx
Notochord and nerve cord
Tail

FIG. 330

INDEX

Abdomen, 220, 234, 238, 240–42, 244–45, 250–53, 258–59, 261
Aboral plate, 271
Acicula, 144, 147, 149–50
Acidiacea, 293
Acoela, 107–8, 110, 112
Acontia, 83, 87
Actacidea, 219
Actin, 194
Actinaria, 49
Actinopodea, 1
Actinosphaerium, 1, 20
Actinula, 52, 54
Adhesive disc, 112–13
Adhesive gland, 116–19
Aequipecten irradians, 173, 190, 196, 204
  dissection and drawing, 197
  SEM, of gill, 205
Albumin gland, 185
  duct of, 185
Alcyonaria, 49, 90
Alveolus, 281
Ambulacral area
  in holothurians, 284
  in irregular urchins, 282
Ambulacral groove, 270
Ambulacral ossicle, 267
Ambulacral plate, 265, 275, 282
Ambulacral section, 272
Ambulacral system, 264
Ambulacrum, 282–83
Amoeba, 1
  drawing, 15
  proteus, 14–16
  SEM, 17
Amoebida, 1, 16
Amoebocyte, 46, 90
Amphiblastula, 46
  drawing, 47
Amphipod, 242
  caprellid, 242
Amphipoda, 219
Amphitrite ornata, 139, 150
Ampulla
  in echinoderms, 265–67, 276, 281, 287–88
  in *Paramecium*, 23
  tentacular, 287
Anal papilla, 217
Anal plate, 174

Anal pore, 100–101
Anal vesicle, 137
Animalia, 33
Anisomyarian, 194
Annelid, 134–35, 146, 174
Annelida, 139
Annulations, 156, 166, 171
Anomura, 219
Anopla, 121
Anostraca, 219, 234
Antenna
  in amphipods, 243
  in barnacles, 238
  in copepods, 236–37
  in crustacean larvae, 261
  in *Cypris*, 235
  in *Daphnia*, 234
  in decapods, 245, 249–51, 257–59
  in isopods, 241
  in ostracods, 234
Antennule, 235–36, 238, 245, 249
Anthocodium, 90–92
Anthomedusae, 49, 52, 56, 60
Anthozoa, 49, 82
Antrum, 113
Anus
  in chordates, 294–95
  in copepods, 236–37
  in crustacean larvae, 261
  in decapods, 245
  in echinoderm larvae, 291
  in echinoderms, 275, 282, 284, 287
  in *Limulus*, 220, 228
  in molluscs, 175, 177, 195, 197, 199, 213
  in nematodes, 127
  in nemertine worms, 123
  in rotifers, 131
Aorta, 185, 190
Aortic arch, 227
Aplysia, 178
Apodeme, 227
Apopyle, 37–39, 43
Appendage
  abdominal, 244, 261
  in annelids, 150
  biramous, 144
  in *Cassiopeia*, 81
  clasping, in *Limulus*, 223

  in copepods, 236
  in ostracods, 234
  raptorial, in *Caprella*, 242
  thoracic, 234, 242–43, 258, 261
  trunk, in *Cypris*, 235
  uniramous, 240
Arachnid, 220, 232
Arbacia punctulata, 263, 273
  dissection, 276–77, 279
  drawing, 275, 281
  pluteus larva, 290
Arbacioida, 263
Arcella, 1
  discoides, 16
  SEM, 17
Arcellinida. *See* Testacida
Archiannelida, 139
Archiannelids, 140
Architeuthis, 210
Arenicola cristata, 139, 156
  dissection and drawing, 157
Arenicolidae, 139, 156
Aristotle's lantern
  musculature, 276
  protractor muscle, 281
  retractor muscle, 281
  skeleton, 276
Arm
  in asteroids, 264
  brachiolar, 290–91
  larval, 291
  in ophiuroids, 270–71
  in squids, 216–17
Artery
  abdominal, 249
  visceral, 179
Arthropod, 224, 230
Arthropoda, 219
Aschelminthes, 121, 126, 130
Ascoglossa, 173
Asterias forbesi, 263–64, 266, 268
  dissection, 265
  drawing, 265
  SEM, of arm, 266
  SEM, of dermal spine, 269
Asteroidea, 263
Asteroids, 264, 270
Astrangia, 49, 92
Atelostomata, 263
Atrium, 294–95

pseudolamellibranch, 206
secondary, in pulmonates, 182
thoracic, 242
Gill filament, 174–75, 200–202, 204, 206, 208
ordinary, 206–7
plicate, 206–7
principal, 206–7
Gill lamella, 220, 222–23
Gill operculum, 220–21, 223
Gill slits, pharyngeal, 294
Gill suspensory membrane, 193, 201
Girdle, 12–13
in chitons, 174–75
Gizzard, 159, 161, 227–229
Glycera dibranchiata, 139, 142
Glyceridae, 139
Gnathobase, 220, 222–23
Gnathobdellida, 139
Gnathopod, in amphipods, 242–43
Gnathostomata, 263
Gonad, 65, 77, 83, 133, 135, 137, 166, 193, 195
in asteroids, 265
in bivalve mantle, 195
in decapods, 254
in holothurians, 284, 287
in squids, 214
Gonium, 1, 6–7
Gonodendra, 62
Gonopalpons, 62–63
Gonophore, 52–53, 56–57, 60–63
Gonopod, 244–45
Gonopore, 113, 164, 283
in *Limulus* male, 222–23
in pycnogonids, 232
Gonotheca, 53
Gonozooid, 50–52, 56, 60–63
Gorgonacea, 49, 92
Gorgonin, 92
Grantia. See Scypha
Green gland, 249
opening, 245
Gut, 112–13, 141, 149
anterior branch, 113
in echinoids, 276
in holothurians, 284
lumen, 163
posterior branch, 113
Gymnostomata, 1
Gyrodinium, 1, 12
SEM, 13

Haplotaxida, 139
Head
in amphipods, 243
in copepods, 236
in *Daphnia,* 234
in molluscs, 176–78, 188
in ostracods, 234
in pycnogonids, 233
in squids, 216

Heart
accessory, 190
in annelids, 161
branchial, 213
in *Ciona,* 295
in *Daphnia,* 235
in decapods, 246, 249, 254, 257
in *Limulus,* 224, 227, 230
in molluscs, 190, 193
in pycnogonids, 232
systemic, 213
Heart urchin, 282
Heliozoa, 1
Heliozoans, 20
Hepatopancreas, 249
Hermaphrodite, 78, 110, 164, 166, 182, 196
protandric, 198
Hermaphroditic duct, 185
Hermit crab, 60
Hydractinia colony on shell, 60
Heterocoela, 33
Heterodonta, 173, 192
Heteronemertea, 121
Heterotrichida, 1
Hexactinellida, 33
Hind gut, 137, 235
Hinge, 188–89, 195–97, 199
in *Limulus,* 220, 230
in ostracod valves, 234
Hinge teeth, 192–93
Hirudinea, 139
Hirudinidae, 139
Hirudo medicinalis, 139, 166, 170
dissection, 167
drawing, 169
Holothuroidea, 263
Holothuroids, 264, 284
Homocoela, 33
Hoploplana, 110
Horn, 261
Horseshoe crab, 112, 220, 230
Hydra, 49, 66, 68, 70, 94
cross section drawing, 69
life cycle, 67
SEM, 71
SEM, of nematocyst, 95
Hydractinia, 49, 60
SEM, 61
Hydranth, 52–53, 56
Hydrocaulus, 53, 57–58
Hydrocoel lobe, 291
Hydroida, 49
Hydroides dianthus, 139, 152
SEM, 153
Hydromedusa, 64
Hydrostatic skeleton, 50, 82, 122, 126, 142, 146, 190
Hydrotheca, 52–55, 58–59
Hydrozoa, 49–50, 72
Hymenostomata, 1
Hypobranchial gland, 177
Hypodermis, 254, 257
Hypostome, 54, 58, 70
Hypotrichida, 1

Ilyanassa obsoleta, 173, 180
Inarticulata, 121
Ink sac, 213
Interambulacral plate, 275
Interambulacrum, 282
Intestine, 291
in annelids, 159, 161–62, 169
anterior, 115
in *Arbacia,* 281, 287
in brachiopods, 133
cavity of, 125
in *Ciona,* 295
in copepods, 237
in decapods, 246, 249
in echiurans, 137
in molluscs, 185, 213
in nematodes, 127, 129
in nemertines, 123
in platyhelminths, 110, 115
in rotifers, 131
in sipunculids, 135
Introvert, retracted, 135
Isomyarian, 188
Isopod, 242
marine, 240
terrestrial, 240
Isopoda, 219
Isthmus, 129

Jassa, 219, 242
Jaws
in annelids, 142–43, 147, 150, 169, 171
in ophiuroids, 270
in squids, 216–17

Kidney, 185, 213
Kinetodesmal fibers, 24
Kinetofragminophora, 1

Labrum, 239
in heart urchins, 282–83
Lappet, 74–75
rhopallial, 77
Larva, 46, 127
actinula, 52, 54
amphiblastula, 39, 46
in arthropods, 220
bipinnaria, 290
brachiolaria, 290
in cephalopods, 216
in crustaceans, 260
in echinoderms, 264, 290
megalopa, 260
mysis, 260
nauplius, 236, 260
parenchymula, 46
planula, 50–51, 72, 74, 94
pluteus, 290
protonymphon of pycnogonids, 232
protozoea, 260
tadpole larva of *Botryllus schlosseri,* 294

301

retractor, 82–83, 87, 135, 163, 185, 284, 287
  stomach retractor, in *Asterias*, 265
Muscle bundle, in ctenophores, 101
Muscle fiber network, in ctenophores, 102–3
Muscle layer, 116
  circular, 163, 266
  intestinal, 163
  longitudinal, 266
Muscular system, 122, 146
Musculature, 116, 268
  ampulla, in *Asterias*, 266
  in bivalve gill filaments, 208
  circular, 118–19
  in decapods, 246
  parapodial, 149
Mya arenaria, 173
Myoida, 173
Myoneme
  in *Hydra*, 68
  in *Vorticella*, 30
Myosin, 194
Mysis, 260
Mytiloida, 173
Mytilus edulis, 173
  dissection and drawing, 194–95
  gill, 202
  SEM, of gill, 203

Nauplius, 260
  larval stage, in copepods, 236
Nautilus, 210
Navanax, 178
Nematocyst, 50, 54, 78, 80, 82, 94, 104
  adhesive, 94
  barbed, 94
  batteries of, 58, 62, 64, 70, 94–95
  discharged, 88
Nematocyst types
  holotrichous isorhiza, 94
  spirocyst, 94
  stenotele, 94
Nematoda, 121, 126, 128
Nematode, 126, 128
Nematomorpha, 194
Nematophores, 58–59
Nematotheca, 59
Nemertinea, 121–22
Neogastropoda, 173
Neopilina galatheae, 174
Nephridium, 135, 137, 141, 147, 149, 163, 169, 171
Nereidae, 139
Nereis virens, 139, 142, 146, 148
  dissection and drawing, 147
  drawing of cross section, 149
  drawing of parapodium, 145
Nerilla antennata, 139–40
  SEM, 141

Nerve cord, 112, 123, 190
  dorsal, 294–95
  longitudinal, 113
  ventral, 118–19, 135, 137, 147, 149, 157, 159, 163, 169, 171, 228–31, 246
Nerve net, 74, 79, 82, 92, 96
  subepidermal, 68
  subgastrodermal, 68
Nerve ring, 127, 129
  in asteroids, 264
  circumoral, in *Limulus*, 228–29
Nervous system, 122
  in annelids, 146
  cephalopod, 210, 216
  in cnidarians, 68
  in echinoderms, 264
  in molluscs, 174, 178, 188, 190, 210
  in platyhelminths, 110, 112
Neuron, 68, 104–5
  bipolar, 96
  multipolar, 96
Neuropodial lobe, 142
Neuropodium, 144–45, 151, 157
Nidamental gland, 178, 215
Notochord, 294–95
Notopodial lobe, 142
Notopodium, 144, 150–51
  aliform, in *Chaetopterus*, 150–51
  branchial lobe, 145
  fan, in *Chaetopterus*, 150–51
Nucleus, 3, 5, 15, 35, 69, 105
Nuculoida, 173
Nutritive cell mass, 109

Obelia, 49–51, 54, 64
  colony drawing, 53
  life cycle, 51
  SEM, 55
Ocellus, 65, 74, 78
  pigment cup, 190
Octocorallia, 49
Ocular plate, 275
Odontophore, 178
Oligochaeta, 139
Oligochaetes, 164
Oligohymenophora, 1
Oniscus asellus, 219, 240
Onuphidae, 139
Ooze
  calcareous, 18
  radiolarian, 20
  siliceous, 20
Operculum, 94
  in molluscs, 177–79, 186–87
  in serpulids, 152–55
Ophioderma brevispina, 263, 270
Ophiurida, 263
Ophiuroidea, 263
Opisthobranchia, 173
Opisthosoma, 220–21, 228
Oral arms, 72, 74–78, 80

Oral disc, 82, 87–88
Oral groove, 23
Oral lobe, 100–102
Oral plate, 271
Oral shield, 270–71
Ornithocercus, 1, 12
  SEM, 13
Osculum, 34–35, 42
Osphradium, 177
Ossicle, 264
  jaw, 268–69
  skeletal, in holothurians, 284
Ostium, 36–37, 39
  in bivalve gills, 206, 208–9
  in crab heart, 257
  in crayfish heart, 249
  in pycnogonid heart, 232
Ostracod, 234
Ostracoda, 219
Ova
  mature, 9–11
  in molluscs, 208
Ovary
  in annelids, 169, 171
  in chordates, 295
  in copepods, 237
  in *Hydra*, 66, 68–69
  in molluscs, 177, 197, 215
  in nematodes, 127, 129
  in platyhelminths, 111
  in rotifers, 131
Oviduct, 111, 127, 237
Oviductal opening, 215
Ovigerous leg, 232–33
Ovotestis, 185
Oyster, 194
  dissection, 198
  gill, 206

Palaeotaxodonta, 173, 192
Pallial groove, 174–75
Palp, 140–43, 147
  in bivalves, 188–90, 195
  labial, 192–93, 197, 199, 202
Palp appendage, 192–93
Pandorina, 1, 6–7
Papilla, adhesive, in tunicate larva, 295
Papula, 264, 270
Paramecium, 1, 22, 24, 26
  conjugation, 25
  drawing, 23
  pellicle, 25
  SEM, 25
  trichocysts, 25
Paramylon body, 2–5
Paramyosin, 194
Parapodium, 140, 142–44, 150, 152, 156, 158, 164
  drawing, 145
Parasite, 108, 112, 126, 238
Parenchyma, 108, 116–18, 122
Parenchymal cells, 119
Parenchymula, larva stage, in sponges, 46
Parthenogenesis, 130

Stigma, 3
Stolidobranchiata, 293
Stomach
    in annelids, 157
    cardiac, in *Asterias,* 265
    cardiac, in crab, 257
    cardiac, in crayfish, 249
    in chordates, 294–95
    in decapods, 254
    in echinoderm larvae, 290–91
    eversion, in asteroids, 264
    in holothurians, 287
    in molluscs, 177, 199
    in nemertine worms, 123
    pyloric, in *Asterias,* 265
    pyloric, in crayfish, 249
    in rotifers, 130–31
Stomach intestine, 147, 228–31
Stomatocoel, 290
Stomodaeum, 237
Stone canal, 266, 284
Strobila, 72–74
Strobilation, 72
Strongylocentrotus droebachiensis, 263, 282
Stylochus, 107, 110
    drawing, 111
Subgenital pit, 80
Sucker
    adhesive, 290
    anterior, 166, 171
    posterior, 166, 169, 171
    in squids, 213, 217
    on tube feet of echinoderms, 264, 266
Sulcus, 12–13
Suranal plate, 275
Swimmeret, 244–45
Symmetry
    bilateral, 88, 108, 264, 282, 284, 290
    biradial, 88, 282
    pentameric radial, 264, 282, 284
Synapse, 104–5
    giant, in squid, 210, 216
Syncytial theory, 108

Tail filament, 109
Tardigrada, 58
Telson, 220, 228, 230, 236–37, 241, 244–45, 249
    flexor of, 227
Temnopleuroida, 263
Tentacle
    in annelids, 140–43, 145, 147, 150–51
    in asteroids, 264
    auricular, 100–101
    capitate, 57–58, 60
    in cnidarians, 50, 52–54, 56–58, 62, 64–65, 68, 70, 74–80, 82, 87–88, 90, 92, 94

in ctenophores, 100–101, 104
filiform, 57
guard, 197
in holothurians, 284
in *Lingula,* 132
mantle, 191
marginal, 78
in molluscs, 177, 185, 190–91, 193, 196, 199, 213
oral, 65
pallial, 190
pinnate, 90, 92
rhopallial, 79
sensory, 197
siphonal, 193
in tunicates, 294–95
Tentacle sheath, 100–101
Tentaculata, 99
Terebellidae, 139, 150
Test, 16, 18, 20
    in sea urchins, 264, 273, 276, 282
Testacida, 1
Testis, 66, 113, 169, 171, 197, 213, 295
Tetrahymena, 1, 28
    SEM, 29
Teuthoidea, 173
Thais lapillus, 173, 180
Theca, 12, 92
Thorax, 150, 234, 236, 238
Thyone briareus, 263, 284, 288
    dissection, 285, 288
    drawing, 287
Toe, 130–31
Toxocysts, 26
Trematoda, 108
Trichocyst, 24, 26
Tricladida, 107, 112, 114
Trilobite, 220
    larva, of *Limulus,* 203
Trochophore, 250
    larval stage, in bivalves, 186
Trochus, 131
Trophi, 130
Tube feet, 264–66, 270, 272, 275, 282, 284, 287
    cavity of, 267
    oral, 275
    SEM, 266
Tubercle, 282
Tubificida, 139
Tubularia, 49, 52, 56
    actinula larva, 54
    SEM, 53
Tubules of Cuvier, 284
Tunic, 294
Tunicate, 294
Turbatrix aceti, 121, 126, 128
    drawing of whole mount, 127
Turbellaria, 107
Turbellarian, 112
    acoel, 108
Typhlosole, 163

Umbo, 188–89, 195, 197
Undulating membrane, 28–29
Urechis caupo, 121, 136
    dissection and drawing, 137
Urochordata, 293
Uropod, 241, 243–45, 249, 258–59
Uterus, 127

Vagina, 127, 169, 185
Valve
    in crustaceans, 234–35
    in molluscs, 180, 186–88, 192, 196, 198, 213
Vas deferens, 169, 171, 213, 257
Vascular ring, 228–29
    drawing, in *Limulus,* 229
Vein, branchial, 213, 257
Velar fold, 196–97
Velar lobe, 186
Veliger, 216
    larva, of bivalves, 186
Velum
    in cnidarians, 64–65
    in molluscs, 186–87
Vena cava
    anterior, 213
    posterior, 213
Veneroida, 173
Ventricle, 185, 189–91, 195, 197, 199
Vessel (blood), 185, 209, 228, 230
    afferent, 201
    branchial, 157, 203, 209
    contractile, 190
    dorsal, 147, 149, 157, 163
    efferent, 201
    lateral, 147
    subneural, 163
    ventral, 147, 149, 159, 163
Vestibule. *See* Reservoir
Vinegar eel, 126
Volvocida, 1, 6
Volvox, 1, 6–7, 10–11
    life cycle, 9–10
Vorticella, 1, 30
    SEM, 31

Walking leg, 220–22, 230–33, 245, 253, 261
    muscles, 227, 231, 257
    thoracic. *See* Periopod
    uniramous, 240
Water bear, 58
    SEM, 59
Water vascular system, 264, 266, 272, 284
Wheel organ, 130

307